IEE HISTORY OF TEC'

Series Editor: Dr B. Bo

History internat broadca

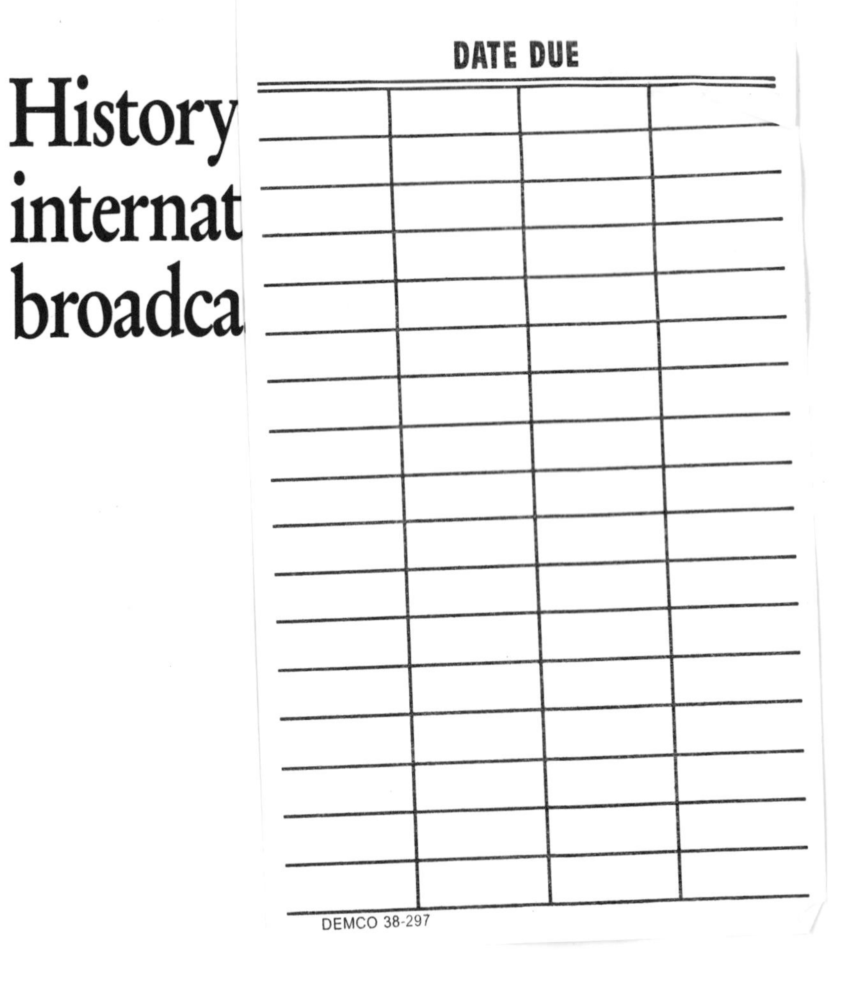

Other volumes in this series:

Volume 1 **Measuring instruments—tools of knowledge** P. H. Sydenham
Volume 2 **Early radio wave detectors** V. J. Phillips
Volume 3 **A history of electric light and power** B. Bowers
Volume 4 **The history of electric wires and cables** R. M. Black
Volume 5 **An early history of electricity supply** J. D. Poulter
Volume 6 **Technical history of the beginnings of radar** S. S. Swords
Volume 7 **British television—the formative years** R. W. Burns
Volume 8 **Hertz and the Maxwellians** J. G. O'Hara and D. W. Pricha
Volume 9 **Vintage telephones of the world** P. Povey and R. A. J. Earl
Volume 10 **The GEC Research Laboratories 1919-84** R. J. Clayton and J. Algar
Volume 11 **Metres to microwaves** E. B. Callick
Volume 12 **A history of the world semiconductor industry** P. R. Morris
Volume 13 **Wireless: the crucial decade, 1924-1934** G. Bussey
Volume 14 **A scientist's war—the war diary of Sir Clifford Paterson 1939-45** R. Clayton and J. Algar (Editors)
Volume 15 **Electrical technology in mining. The dawn of a new age** A. V. Jones and R. P. Tarkenter
Volume 16 **'Curiosity Perfectly Satisfyed:' Faraday's travels in Europe 1813-1815** B. Bowers and L. Symons (Editors)
Volume 17 **Michael Faraday's 'Chemical notes, hints, suggestions and objects of pursuit' of 1822** R. D. Tweney and D. Gooding (Editors)
Volume 18 **Lord Kelvin: his influence on electrical measurements and units** P. Tunbridge

Front cover: By kind permission of Thomson-CSF, France. Scale model of ALLISS (see Appendix III), a new concept in shortwave global broadcasting; 17 x 500 kW ALLISS will enter service with *Radio France International* by 1997.

History of international broadcasting

James Wood

Peter Peregrinus Ltd. in association with
The Science Museum, London

Published by: Peter Peregrinus Ltd., London, United Kingdom

Reprinted 1994
Paperback edition 1994

Peter Peregrinus Ltd.,
The Institution of Electrical Engineers,
Michael Faraday House,
Six Hills Way, Stevenage,
Herts. SG1 2AY, United Kingdom

British Library Cataloguing in Publication Data

A CIP catalogue record for this book
is available from the British Library

ISBN 0 86341 302 1 (paperback)

ISBN 0 86341 281 5 (casebound)

Printed in England by Short Run Press Ltd., Exeter

'Communication systems are neutral. They have neither conscience nor morality; only history. They will broadcast truth or falsehood with equal facility. Man communicating with man poses not a problem of how to say it, but more fundamentally what he is to say'

Edward R. Murrow
Wartime broadcaster, Director US Information Agency
Voice of America 1961–1964 Died 30 April 1965

Contents

Preface

In attempting a history of international broadcasting in a single volume it was inevitable that much interesting material would have to be discarded; I do not have the space to describe the 160 or so broadcasting authorities around the world.

Of the major broadcasters, I have covered Voice of America, because its transmission network is the biggest in the world; the BBC World Service, because it commands the greatest listening audience in the world; Deutsche Welle, for its political influence and role at the centre of east-west relations; and Radio France International, for its transmission technology and its programme culture of projecting France to the world.

Of the major commercial international broadcasters, I have mentioned Radio Monte Carlo for its unusual history and its present-day super-power broadcasting capability on all wavebands. Among religious non-government broadcasting organisations, I have included data on the two most influential and respected: Christian Science Monitor, with its subsidiary Herald Broadcasting, and Trans-World Radio.

I have been unable to include many other eminent broadcasting authorities, like Radio Canada and Swiss Radio International, to name but two; to these I offer apologies. My second regret is not including descriptions of studio programming. I have concentrated on transmission, first because this is where my experience lies—as a transmitter designer—and secondly because it is an aspect of broadcasting that receives less publicity than the studio side.

In researching this book I met with co-operation, kindness and often generosity in arranging visits to broadcasting authorities and the plants of the broadcasting-equipment manufacturers.

My personal thanks go to the individuals in many organisations who went out of their way by assisting wherever possible: George M. Badger (EIMAC), Patrick Bureau and Guy-Noel Le Carvennec (Thomson-CSF), J.A. Cant and David C. Lees (BBC), D. Adams (TWR), Johannes Denteler (Siemens Tubes), Robert E. Frese (Director of Operations, VOA), Jürgen Graaff (Managing Director, AEG Telefunken), Dr. Ernst Hoene (Head of Development, Siemens Tubes), Gunter Haufe (RIAS Berlin), Charles Kalfon (Director, Thomson Tubes Electroniques), Hans Leupold (AEG Telefunken), Raymond Rowe (Marconi), Bernard Poizat (RMC), Martha B. Rapp and Robert R. Weirather (Harris), David M. Russell

(Continental), Jukka Vermasvuori (YLE), Dr. Wolfram Schminke (ABB), Peter Wildish (GEC-Marconi), Robert Wilson (Int. Press, BBC World Service) and Thomas Yingst (Vice President, Harris Broadcast).

I have drawn on articles published in *International Broadcasting* in my capacity as Transmission Correspondent, and I would like to express my thanks to the publishers concerned. To the authors and publishers of the many references sources in the bibliography I extend thanks. In particular I would single out the works of Barnouw, Briggs, Boyd, Duus, Herzstein, Head & Sterling, Nichols, Pawley, Sperber and former President Nixon.

The contents of this book have been supplemented where appropriate from memory, personal experiences and personal archives. Finally, I would like to thank my publishers, Peter Peregrinus Ltd., and the IEE for giving me the opportunity of writing this history. My personal thanks go to John St Aubyn and the Series Editor, Dr. Brian Bowers, for encouragement and support and to Mike Kew for his hard work in editing. I am indebted to Fiona MacDonald, Production Editor, for her advice and assistance.

James Wood
1992

Acronyms and abbreviations

ABB	Asea Brown Boveri
ABC	American Broadcasting Company; Australian Broadcasting Corporation
ABSIE	American Broadcasting Service in Europe
AEG	German Electric Company
AFB	American Forces Broadcasting
AFRS	American Forces Radio Service
AM	Amplitude modulation
AT&T	American Telephone & Telegraph
AWA	Amalgamated Wireless Australasia
AWR	Adventist World Radio
BBC	British Broadcasting Corporation (Company before January 1927); Brown Boveri Company (became Asea Brown Boveri in 1989)
BEF	British Expeditionary Force
BTH	British Thomson Houston
CBC	Canadian Broadcasting Corporation
CBS	Columbia Broadcasting System
CIA	Central Intelligence Agency
CRC	Crosley Radio Corporation
DAM	Dynamic amplitude modulation
DCC	Dynamic carrier control
DR	Deutsche Rundfunk
DSB	Double sideband modulation
DW	Deutsche Welle
FBIS	Foreign Broadcasts Intelligence Service
FCC	Federal Communications Commission
FCO	Foreign & Commonwealth Office
FEBA	Far East Broadcasting Association
FM	Frequency modulation
FO	Foreign Office
GE	General Electric
GEC	General Electric Company
GPL	Gau Propaganda Central Office

HF	High frequency
HMGCC	H.M. Government's Communication Centre
IRIB	Islamic Republic of Iran Broadcasting
ITU	International Telecommunications Union
LW	Long wave
MOI	Ministry of Information
MUSA	Multiple unit steerable array
MW	Medium wave
NAB	National Association of Broadcasters
NBC	National Broadcasting Company
NHK	Nipon Hoso Kyokai (Japanese Broadcasting Corporation)
NRB	National Religious Broadcasters
OTH	Over the horizon (broadcasting)
ORF	Osterreichischer Rundfunk
OSE	Overseas extension
OSS	Office of Secret Service
OWI	Office of War Information
PDM	Pulse-duration modulation
PSM	Pulse step modulation
PTT	Postal, Telephone and Telegraph
PWD	Psychological Warfare Division
PWE	Political Warfare Executive
RBI	Radio Berlin International
RCA	Radio Corporation of America
RF	Radio France
RFE/RL	Radio Free Europe/Radio Liberty
RFI	Radio France International
RIAS	Rundfunk im Amerikanischen Sektor
RIZ	Radio Industries Zagreb
RL	Radio Luxembourg
RTL	Radio Tele Luxembourg
SOE	Special Operations Executive
SRI	Swiss Radio International
SSB	Single sideband
SW	Short wave
TDF	TeleDiffusion de France
TWR	Trans-World Radio
UAE	United Arab Emirates
UHF	Ultra-high frequency
USA	United States of America
USIA	United States Information Agency
USSR	Union of Soviet Socialist Republics
VHF	Very high frequency
VOA	Voice of America
VOFC	Voice of Free China
WARC	World Administrative Radio Conference
WBC	Worldwide Broadcasting Corporation
WE	Western Electric
WRTV	World Radio & TV Handbook
YLE	Yleisradio (Finnish Broadcasting Company)

Acknowledgments

I am grateful to the following:

Publishers Michael Joseph for permission to quote passages from 'EDWARD MURROW: his life and times.' Ann SPERBER (1987)

Oxford University Press, NY, for permission to use background material from 'A history of broadcasting in the US' Three volumes: Vol. 1 'A tower in Babel' BARNOUW (1966); Vol. 2 'The Golden Web' BARNOUW (1968); Vol. 3 'The image makers' BARNOUW (1970)

Warner Books Inc., NY, for permission to take short extracts from pp. 41, 251, 286, 313, 338, 341 'The real war' Richard NIXON (1980)

BBC Enterprises for permission to use material from pp. 348, 351 'BBC Engineering 1992–1972' Edward PAWLEY (1972)

Publishers W.H. Allen for permission to draw on material from 'Radio Luxembourg' R. NICHOLS (1983)

Oxford University Press for permission to use material from 'The BBC: the first 50 years' Asa BRIGGS (1985)

Temple University Press, PA, USA, for permission to use material from: 'Broadcasting in the Arab World' Douglas A. BOYD (1982)

The Institute of Electrical and Electronic Engineers, NJ, for permission to use material from papers published in *IEEE Transactions* Vol. 34 No. 2 June 1988
'Jamming in the broadcast bands' SOWERS HANDS RUSH pp. 108
'The renaissance of private SW broadcast stations' JACOBS pp. 87–93
'The early history of US international broadcasting' WELDON pp. 82–86
'Evaluation of 500 kW SW transmitters' BERMAN *et al.* pp. 147–153

Verso Publications and NLB for permission to take quote from 'The making of the second cold war' Fred HALLIDAY (1983) p. 4

Publishers Hamish Hamilton for the use of extracts from 'The war that Hitler won' R.E. HERTZSTEIN (1979). Every attempt was made to clear permission and the publisher would be very happy to hear from the copyright owner whom we were unable to trace.

Publishers Kodansha International for the use of material drawn from 'Tokyo Rose: Orphan of the Pacific' Masayo DUUS (1979). Every attempt was made to clear permission and the publisher would be very happy to hear from the copyright owner whom we were unable to trace.

To the following manufacturers of broadcast transmission systems for the use of published technical data and customer sales references of high power transmitters from the 1950s to present day.

AEG Telefunken, Berlin, Germany
Asea Brown Boveri ABB, Baden, Switzerland
Continental Electronics, Dallas, Texas, USA
Marconi Communication Systems, Chelmsford, UK
Radio Industries Zagreb RIA, Zagreb, Croatia
Thomson CSF, Gennevilliers, France

To International Broadcasting Authorities BBC World Service, Deutsche Welle, Christian Science Monitor, Radio France International, RIAS Berlin, Radio Monte Carlo, Trans World Radio, YLE Finnish Broadcasting, Voice of America and others for assistance in providing technical data and in some cases the arranging of visits to transmitting stations.

List of illustrations and their sources

Introduction: Growth in information broadcasting

Every new medium of information has made advances on the previous generation, and in so doing has established new values and created an awareness of increased power. National radio broadcasting, in its early days in the 1920s, was no exception to this rule; but it was the discovery of the short waves and the subsequent development of the so-called super transmitters some five decades later that permitted radio to eclipse all other forms of mass media with its awesome potential.

Communications media have been evolving since the time of the pharaohs. In ancient Egypt, the first communications revolution came with the transition from stone to a parchment produced from the papyrus plant as the medium for recording information. The portability of papyrus and its ease of handling greatly assisted the exercise of monarchial power. The new medium also created an elite corps of scribes, who became a powerful group in their own right.

The invention of printing in Europe was responsible for a revolution in learning. Ample supplies of paper encouraged the spread of knowledge and stimulated commerce. At the same time the position of the church as the sole possessors of knowledge was undermined, and heretical ideas about nature and the universe rapidly spread through Europe. Communications became central to political life; governments survived or fell by their mastery of information.

After printing, the next revolution came with the invention of the telegraph: the wire age (1844–1900) culminated in the establishment of the first international and multinational communications companies. The international cable telegraph companies were now possessors of power. This new medium enabled the powerful nations like Great Britain to maintain contact with their far-flung colonies. Then came wireless telegraphy (1901–1926). Neither this nor the telegraph was seen as a threat by the newspaper owners, but rather as an aid to the printed word. It gave the opportunity to transmit news instantly, and brought about the syndication of news from press agencies.

However, when the invention of the radio tube brought about possibilities of sound broadcasting of music and news, the press barons saw things in a different light. As soon as a tiny station in Pittsburgh broadcast the result of the American presidential election of 1920, newspaper proprietors were not slow to see the threat. In Britain in 1922 the press barons saw to it that the newly formed British Broadcasting Company should not be permitted to broadcast news. Not until the

BBC became a Corporation, with a Royal Charter, was it permitted to broadcast news.

When it became obvious that sound broadcasting was here to stay, newspaper owners in America started buying radio stations to extend their control over media. At this time, in the mid-1920s, the transmission range of radio broadcasting was limited by technology: the high-power triode had yet to appear, and as a consequence transmitter output power was measured in hundreds of watts rather than in kilowatts. Limited to such powers, the useful range of a broadcast varied from 50 to 100 miles.

The discovery of the short waves opened up the possibility of global broadcasting, although with no certainty of performance and reliability of service. In America, short-wave broadcasting became the means of extending the range of entertainment broadcasting to more distant parts (mainly Central America); but in Europe it was developing on entirely different lines. At first it was perceived as the means of broadcasting to distant colonies, such as Australia and Canada, but in 1936 it took a different line. This was the time the BBC began propaganda broadcasts to the Arab world, partly to counteract similar broadcasts from Italy, which also had colonial interests in the region. This date may be regarded as initiating propaganda broadcasting with two players, but by 1939 Germany and Russia were also honing the techniques of the new propaganda. Today the number of players has risen to over 100 countries.

It was the onset of the Second World War that brought about the first explosion of international propaganda broadcasting; it was a powerful weapon of war for all participants, but most of all for the Allies in a subversive role, creating resistance, havoc and confusion in the occupied countries, and later, after the invasion of Europe, as an instrument for creating stark terror in Germany.

Far from dying with the end of the war, propaganda broadcasting continued to expand apace. The Cold War, in Europe and in much of the rest of the world, was fought mainly with words. The Nazi threat had been replaced by communism. Propaganda broadcasting again proved to be an awesome weapon, eclipsing its performance up to 1945 and playing a dangerous role in the war of ascending oscillatory antagonism between East and West.

Growth in international broadcasting has been explosive. The number of countries investing in the medium rose from four in 1939 to well over 100 by the 1980s. At the same time, transmitter output powers have risen from 10 kW in 1936 to 500 kW by 1985 and even some 1000 kW by 1989. Equally impressive has been the growth in sheer numbers of transmitters around the world.

Wars, whether hot or cold, act as a spur to the advancement of many technologies; money can always be found for weapons of war. By the end of the Second World War Britain was the world's largest propaganda broadcaster; its output ran to 850 programme-hours per week in 46 different languages. This was greater than the total output from the USA and the USSR combined. In 1950, when Britain was crippled with massive war debts, the government cut this figure back to 643 hours per week, but even this figure exceeded that of the two superpowers. Some 40 years later Britain has fallen to fifth place in the world league table of international broadcasters, being overtaken by the USA, USSR, the People's Republic of China and (West) Germany.

Transmission technology has made great strides. By 1943 the 100 kW SW transmitter became a reality. In 1952 the French company Thomson built the first 250 kW transmitter and in 1972 the first 500 kW SW transmitter using a single output tube. Ten years later AEG took the technology revolution one stage further with its high-efficienty pulse-duration-modulation system. In 1985 the Swiss-based company Brown Boveri introduced its unique pulse-step modulation, offering much higher conversion efficiency: an important feature because of the vast amounts of energy associated with running large transmitter complexes. From the efforts of such companies Europe is now the world leader in super transmitter technology. These super transmitters, operating into very-high-gain directive curtain arrays, provide the means for nations to project their broadcasts to every corner of the globe, particularly into the USSR and China, which because of their huge land mass depend on SW frequencies for some of their national radio broadcasting.

Dissemination of propaganda by radio is a highly specialised field; a combination of art and science, whose impact on the listener should not be underestimated. Oral communication is a highly complex process: a combination of many elements in the sending and receiving processes, and of human perception.

The human auditory system can correct for distortion, substitute any missing phoneme, reject information and invent information. Thus perception of speech is an active cognitive process, in which aspects of context merge with phonology. Speed of delivery, syllabic rate, and quality of articulation play vital roles in rendering intelligible speech. Of course, the radio transmitter is not a perfect medium and its limitations have to be reckoned with.

The impact of the spoken word can be enhanced by emotion and rhetoric. Equally, it can be de-emphasised by reading selected passages at a faster rate, or by fast run-on techniques when switching to a different topic. Both devices have the effect of reducing the importance of the first passage in the minds of listeners. All broadcasting authorities and broadcasters are familiar with these techniques.

Propaganda broadcasting has undergone much metamorphosis since 1936. From the harsh, accusatory tones of the late 30s it moved towards a more covert operation in the Second World War. During the Cold War the West used it as a tool to roll back the frontiers of communism. The USSR countered with the only weapon at its disposal; to jam the broadcasts calculated to cause unrest. The 1980s saw both sides winding down the harsh invectives and bellicose nature of the broadcasts. Today most of the 100 or so countries use it as a more refined instrument of foreign policy. Nevertheless, it has to be said that the nations that go to great lengths to expose corruption in other countries will go just as far to conceal information from their own people: one recent example is the Spycatcher trials. It is significant that Voice of America is not permitted to broadcast within the USA, by the terms of its authority.

The political power of radio has done nothing to discourage propaganda broadcasting of a different genre: religious broadcasting. Religion as a tool of politics is generally credited with having created the circumstances that brought about the dismantling of communism in Romania and other Eastern European countries, and more recently in the USSR itself.

The period from the 1980s was the decade of audibility. It signalled the race by many countries to acquire a super-power broadcasting capability, to create a stronger signal on the high-frequency spectrum 3–30 MHz. The trend was set by such broadcasters as the BBC World Service, Radio France International,

Deutsche Welle and ORF Austria. In 1984 the Reagan administration declared that it would not compromise its aim to maintain a free flow of information across territorial and international borders, and made plans for the greatest expansion in the history of international broadcasting.

But, notwithstanding the historical supremacy of Europe and the USA, the greatest expansion in propaganda broadcasting capability is now happening in the Middle East. This is an area of the world where the spoken word dominates the written. Linked by religion, culture and a common language broadcasting has assumed a high degree of importance throughout the Arab world. The Arabic culture is not only an orally based culture; it is also a rhetorical and emotional language. Educated Arab leaders were quick to realise the strategic value of radio broadcasting as a political tool, ideally suited to rhetorical speeches, and because of the common tongue, powerful super transmitters could broadcast the same programme to neighbouring countries.

Arab countries have now entered phase three of their expansion: the acquisition of multi-frequency, global broadcasting capability, to give them a greater importance in the rest of the world and extend the voice of Arabism. Even now some Middle East countries are bringing up on power some of the most powerful broadcasting centres in the world, and are planning to overtake Western powers in terms of modern technology and gross output power. One Arab country has already achieved this goal.

In terms of commanding global audiences, the Middle East countries have a long way to go before they can touch the BBC World Service, for which a recent global analysis showed that it commands a world audience in 90 countries totalling 120 million. This represents an increase of 60% over the estimated total of 75 million for 1980. But then the BBC has had 60 years of unrivalled experience in honing the techniques of propaganda broadcasting; its key to success lies in its image as an independent voice in the world, and not the voice of the British government.

International broadcasting is a highly competitive business conducted by governments through their broadcasting agencies. Propaganda, like any marketable commodity, has to respond to changing markets, and these markets are complex, involving world politics, sudden and unpredictable changes in power, shifts in listening habits and disasters. Broadcasters must take notice of what other broadcasting agencies are saying. Finally, they must always seek to take advantage of evolutionary changes in broadcasting technology.

Today's sixth-generation satellites have revolutionised all aspects of broadcasting and in the case of international broadcasting in the HF spectrum their effect has been far-reaching. The global broadcasting capability of Voice of America, the world's biggest propaganda broadcaster, relies heavily on the use of satellite links to convey its programmes from the studios in Washington to its HF relay network in 'host' countries around the world.

Propaganda broadcasting does yield far-reaching results though often over a long term and not always in a predictable place. The fall of communism in the USSR and Eastern Europe followed four decades of effort by the BBC, VOA and the quasi-covert activities of others such as Radio Free Europe and Radio Liberty; but even so, when the end of communism came in Eastern Europe it was sudden, violent and unexpected by the West.

Part 1: Birth of a medium

Chapter 1

The triode and radio telephony

By 1908, long-wave wireless telegraphy had asserted itself as the most important development of the early 20th century. It held out the promise that it would take over the role of cable communications and make redundant the vast network of cables that spanned the globe. It also offered the potential of becoming a strategic tool in times of war, enabling Britain to maintain communications with those remote outposts not connected to cable.

Moreover, wireless communications offered the prospect to commerce of a cheaper tariff than that charged by the cable companies. At no time was wireless thought to represent a threat to the telephone companies in America and Europe. This complacency on the part of the telephone companies had two reasons: these companies enjoyed a monopoly in this form of communication, and secondly, there seemed little prospect of wireless telephony ever becoming a viable proposition. Apart from a few lone voices, no one in authority had ever talked of a need for telephony broadcasting.

A possible reason for this may have been the misconceptions born of ignorance that held sway in the early years of wireless: the mechanism of wave propagation was not understood, and many people thought that wireless waves travelled as a narrow beam of energy. Others held that broadcasting was actively wasteful: in 1899, *The Electrician* said: 'Messages scattered broadside only waste energy by travelling towards celestial space. They do positive mischief by interrupting the everyday business of wireless stations in the vicinity.'

There were other reasons for the seeming lack of interest in broadcasting. The telephone companies enjoyed the monopoly they held, not only in telephonic communications but also in music broadcasting. In America telephony had become established as a means of transmitting a concert over a telephone line to another audience seated in a theatre, with the output amplified through loudspeakers.

However there were a few who had thought of adapting wireless as a carrier of the human voice and phonograph records. Since 1900, the year when the arc transmitter was introduced, various experimenters had been investigating possibilities of modulating the continuous wave from the arc transmitter, notably Poulsen, Koepel, de Forest and Fessenden.

There were two main obstacles to overcome: how to modulate the continuous wave output from the arc transmitter with sound waves, and how to rectify or demodulate the output from the wireless receiver. In the latter case, it was the discovery of the solid substance carborundum that provided the solution to the problem. General Dunwoodie, of the U.S. Army, was the first person to promote awareness to its unique quality: the ability to rectify very tiny signals.

From 1906—the year of Dunwoodie's achievement—followed the first of the crystal sets that were later manufactured in hundreds of thousands in America and Europe. This marked the first major milestone in receiver design. Although a number of attempts had been made to modulate the arc transmitter, these were unsatisfactory. The real breakthrough came when de Forest invented the audion.

The forerunner of the audion, or triode, was the diode. J.A. Fleming is credited as the inventor of the diode in 1904. He was working at that period as an advisor to the Marconi Company in England. Patents for the diode were taken out in Britain, Germany and the USA, with the object of locking up of the wireless market in the hands of the Marconi company.

However, it is possible to trace Fleming's work back further than the invention of the diode. The first discovery of the phenomenon of electronic emission was made in the USA in 1880 by Thomas Alvar Edison, the inventor of the electric filament lamp. During a series of experiments Edison observed the blackening effect on the inside of the glass bulb, which he correctly deduced was due to electron emission.

To put this theory to the test Edison had an experimental lamp constructed, in which a plate was assembled and connected to an external connection. His theory was correct, and in 1884—more than a decade before the invention of wireless—Edison took out a patent for this device, described as an electrical indicator.

Fleming at this time was employed as an advisor to the Edison & Swan Electric Light Company of England. In 1885 Fleming visited Edison and was shown the result of his research. Much later, after his return to England, Fleming began to experiment with electric light bulbs and managed to produce similar effects to those found by Edison. By 1901 Fleming had accepted an invitation to become a scientific advisor to the Marconi company, working on the development of wireless-telegraphy equipment in connection with Marconi's attempt to send wireless signals across the Atlantic.

It was his work on this project that gave Fleming the inspiration for the diode valve, which was duly patented by the Marconi company in 1904. However, there can be little doubt that the germ of the idea came from his experiments with emissions from electric-lamp bulbs, and the diode valve was the direct descendant of the electric lamp.

While Fleming in England was following this line of research, de Forest in America was diligently pursuing his own line of research. Dr de Forest earned his PhD in 1899—only three years after Marconi had taken out the world's first patent in wireless telegraphy. His doctoral thesis was entitled 'Reflection of Hertzian waves from the ends of parallel wires'.

Although he worked for Westinghouse in the telephone department, his ambitions were on a higher plane: he wanted to adapt wireless to transmitting voice and music. In his diary he wrote: 'what finer task than to transfer the sound of a voice or song to one a thousand miles away'.

This gives a clue to the wisdom at that time, that wireless was a device for communicating with a single place rather than one capable of broadcasting in an omnidirectional manner.

In 1905, just one year after Fleming had invented the diode, de Forest took out his patent for the audion. In effect he had taken the diode and had inserted a third electrode: the grid. Without wishing to detract from the importance of the diode, it was the audion—or triode—that revolutionised wireless, made it possible for transmitters to be modulated, for weak radio signals to be amplified, and laid the cornerstone of the giant radio industry that was to emerge.

In the following year de Forest took out another patent for an improved design of audion. This patent caused a sharp frost in the commercial relations between America and Britain, which up to this time had been of a most cordial fashion. The Marconi company in England had thought that it had the entire world market sewn up with its patent for Fleming's diode. The basis for its litigation in the American law courts was that de Forest's triode was nothing more than an improvement to its patent. The defending American lawyers sought to prove that the triode was to be considered as a totally new device. The results were disastrous for both Marconi and de Forest, as court action was followed by counter actions.

Quite apart from the financial losses incurred in court actions there was another serious consequence for Marconi and for British commercial influence in America, which had historically been good. This was the American decision to break away from British influence by the formation of the Radio Corporation of America (RCA). This American corporation formed from the General Electric company (GE), and became the largest radio corporation in the world.

Fleming—later Sir Ambrose Fleming—never forgave the Americans for their court actions. Fleming had achieved success in many of his discoveries, and his success may have sparked off a streak of vanity. In 1925 he wrote: 'The importance of the diode is also shown by the fact that determined efforts were made by American wireless men to claim the invention for themselves and deprive the writer of the credit for it by re-christening identically the same invention by strange names such as Audion or triode.'

This opinion is not universal, and there are those who regard the true inventor to have been de Forest. Like most inventions, the triode was far from perfect to begin with, and the first samples of the audion were no exception. Early samples were fragile, inclined to be erratic and of variable performance. Given the state of the art of manufacture this is not surprising; the triodes were hand-made to begin with. Other causes of poor performance were of a more fundamental nature: the principles of signal amplification were not properly understood. Nevertheless de Forest had laid down the basis for an industry of the future, which began with research into vacuum-tube theory by such as Langmuir, Carson, Dushman, Van derBijl and others. By chance it was an industry that came to be dominated by Britain: in the mid-1930s, there were no fewer than 65 brand names made or marketed in Britain. But this success was short-lived, and by the late 1940s the numbers had shrunken to half a dozen when the American tube industries such as EIMAC began to dominate world markets.

Ranking in equal importance to de Forest as one of the early pioneers who introduced broadcasting was Reginald Fessenden. Professor Fessenden was a Canadian inventor, generally credited as the first person ever to carry out a radio broadcast. As early as 1901 Fessenden had connected a telephone instrument to a spark transmitter, although without success because the technology was not available. Later he formed his own company to exploit the possibility of sound broadcasting: the National Electric Signaling Company.

Figure 1.1 *Family of modern low and medium power transmitting tubes by Thomson-CSF*

Fessenden was the first to think up an entirely new and even revolutionary approach to the design of continuous-wave radio transmitters: the idea of using rotary alternating-current generators designed to operate at relatively high frequencies. A contract was placed with GE for such a machine, and the development task was assigned to a Swedish-born immigrant by the name of Ernst F.W. Alexanderson, a one-time pupil of Professor Slaby in Berlin. Many at that time thought the idea crazy, yet the Alexanderson alternator went on to become the most important development of the mid 1920s as a revolutionary step forward in long-wave, super-power telegraph transmitters.

Using the first such alternator adapted for telephony transmission, installed at his premises at Brant Rock, Massachusetts, Fessenden went on air on Christmas Eve 1906. Astonished ships' operators, accustomed to the bleeps of morse code, heard the first ever human voice from their receivers; to them it was uncanny. This was followed by a short concert, with a violin solo played by Fessenden himself, and a phonograph recording of Handel's Largo. At the end of the programme Fessenden announced: 'If anyone hears this would they please write to Mr Fessenden at Brant Rock.'

It was heard, all right, by several ships that had been fitted out with experimental wireless equipment. What the broadcast did achieve was to encourage shipping lines to invest in this new technology, which gave the possibility of using voice as an alternative to dots and dashes.

Two years later Fessenden again made radio history by being the first person to broadcast from the top of the Eiffel Tower, earning even greater publicity. But the concept of using telephony broadcasting as an instrument for social entertainment was still a long way off. The next major step came in 1915: the first international broadcast, carrying a human voice a distance of 3000 miles.

The American Telephone & Telegraph company (AT&T), in co-operation with

Standard Electric, built an experimental telephony transmitter and installed it in Arlington, Virginia, using 300 tubes in a parallel arrangement to generate 3 kW of radio frequency power. The signal was received by a station installed at the top of the Eiffel Tower. This was in effect a double milestone; it was the first-ever voice transmission across the Atlantic, and it was the first time that a large number of tubes had been used in a combining arrangement. This same technique was recreated in 1989, when several hundred transistors were combined to generate large amounts of power at UHF.

From this first example in 1915, such was the pace of progress in this new technology that in 1917, when America entered the First World War, GE received a contract from the U.S. Army Signal Corps for the supply of 80 000 radio tubes, commencing at the rate of 500 per week, eventually rising to the staggering figure of 6000 per week towards the end of the contract.

Chapter 2

Origins of entertainment broadcasting

The development of wireless as a medium for broadcasting concerts and phonograph records for the purpose of entertainment was a very slow process. Some of the reasons for this were connected with the lack of the necessary technology, yet when the technology did become available there was still a great lack of awareness of the potential it offered.

Although de Forest is credited as being the father of the device that made broadcasting possible, the father of entertainment broadcasting is David Sarnoff. Sarnoff's career in broadcasting began as an office boy at the American Marconi Company; he rose to office manager, achieved instant fame as a first-class telegraphist by staying at the telegraph key for 72 hours working the stricken *SS Titanic*, and eventually became president of the giant Radio Corporation of America.

The story of Sarnoff's long range vision of the potential of entertainment broadcasting goes back to 1916, when he was contracts manager for the American Marconi Company. The technology was available for sound broadcasting, yet nobody then seems to have realised what broadcasting could mean to the general public.

Sarnoff seems to have been the first to see the potential it offered. He put his proposal in the form of a business plan to his superior, Edward J. Nally, vice president and general manager: 'I have in mind a plan of development which would make radio a household utility in the same sense as the piano or phonograph. The receiver can be designed in the form of a simple music box and arranged for several wavelengths. . . all of which can be neatly mounted in one box. This box can be placed in the parlor or living room, the switch set accordingly and the transmission of music received.' Sarnoff went on to add: 'The same principle can be extended to numerous other fields—as for example—receiving a lecture at home which can be made perfectly audible.'

With the benefit of hindsight it is apparent that Sarnoff could not have expressed his ideas more clearly, and yet Nally is said to have thought the idea 'harebrained'. Sarnoff kept his file copy and bided his time. With the formation of RCA he apparently thought his time had come; in January 1920 he spoke to Owen D. Young and re-submitted his plans in greater detail.

Sarnoff's calculations were based on sales of 100 000 music boxes in the first year, rising to 600 000 in the third year. These estimates earned him a reputation as a prophet; his first year's calculations were right, as were the second year's but by the third year there was a runaway in sales.

The first person in America to broadcast entertainment programmes on a regular basis was Dr Frank Conrad. Conrad was working as a development engineer with Westinghouse in Pittsburgh. While operating an experimental radio-telephone transmitter, in connection with his work with Westinghouse, he fell into the habit of transmitting phonograph records and sports results from Pittsburgh in reply to requests from other radio amateurs. These informal and very much experimental transmissions built up so much interest that they began to get mentions in local newspapers. Conrad's radio station was not unique; indeed, there were several other amateur radio stations doing the same thing, and in Europe a similar thing was taking place. What made the radio station at Pittsburgh unique was the chain of events it had set in motion.

Horne's department store in Pittsburgh, noting the growing interest in Conrad's activities, sensed the commercial opportunity for a new product line. At this time the public did not have receiving sets; as in Europe, receiving sets were still in the experimental stage, and were used only by enthusiasts.

With an astute sense of timing, Horne's store in Pittsburgh ran an advertisement in the local newspaper: 'Concerts picked up by radio here. Amateur sets made available by the manufacturer of the one in our store.'

Westinghouse Electric, seeing the advertisement, saw the possibility of a commercial tie-up with Horne's by making the kits. The vice-president of Westinghouse, realising the need to promote the product, instructed Conrad to modify a second telegraph transmitter to broadcast sound; thus a second radio transmitter in Pittsburgh went on-air in November 1920 from an improvised studio on the roof of the Westinghouse factory in east Pittsburgh. The opening transmissions were timed to cash in on interest in the Harding-Cox presidential election.

On 27 October 1920 the Department of Commerce rushed through a transmitting licence to Westinghouse in response to a formal application six weeks earlier. The call-letters KDKA were assigned, with the authorised use of the wavelength 360 metres. The transmitting installation was completed with only a few days to spare before the presidential election date, and a telephone hook-up had been made with the *Pittsburgh Post* to pass on results as they came into the wire services room at the newspaper.

As a back-up transmitter station, Conrad stayed on duty at the first experimental transmitter in case problems developed, but in the event all went well. KDKA broadcast the results right up till midnight, by which time the final outcome was known with Harding the new president. The listeners in the town went wild with the excitement of the election and the radio broadcasting, which had also made history.

From that night on, entertainment broadcasting snowballed in the USA. KDKA Pittsburgh had set in motion a chain of events that was destined to create a radio entertainment industry without parallel anywhere else in the world, and which would one day dominate the world in terms of its volume production and lead in many technical innovations. KDKA earned itself a place in the history of broadcasting for several reasons: it was one of the first to transmit programmes for entertainment, it was the first radio station to broadcast news to the public, and it was the first radio station licensed to do all these things.

KDKA's fame travelled well beyond America. Over the next decade, as transmitter powers rose, KDKA became the radio station known to all radio enthusiasts in Europe. Probably because of its location on the Eastern Seaboard of the USA, KDKA's signals could be heard regularly in Britain and parts of mainland Europe.

Strangely, although David Sarnoff foresaw the coming of entertainment broadcasting, and alerted both Marconi and later RCA to the potential it offered, RCA failed to act quickly in setting up production lines for radio receivers. The first American company to achieve this was its rival, Westinghouse Electric.

1920–21 saw the birth of American entertainment broadcasting, with the emphasis on entertainment. Commercial broadcasting was to come later. At first the powerful corporations (RCA, General Electric, AT&T and Western Electric) used their radio stations mainly to promote the image of the company—and, of course, there was the added bonus that owning radio stations helped to stimulate the demand for radio receivers and other electrical goods. To these electrical giants, the radio station became a flagship of the company, programmes were not interrupted by advertising and programme quality was high.

The issuing of broadcasting licences began as a trickle in 1920–21 and grew into a hundred in two years, including WOR Newark, WWJ Detroit, WGY Schenectady, WDAF Kansas City, WJZ Chicago and WFAV Nebraska. City after city began the scramble for broadcasting licences.

The flagship of AT&T was station WEAF. It had the advantage that AT&T ran all the telephone communications in the USA, so WEAF was able to make good use of telephone lines and excelled in outside broadcasting. In 1922, it also became the first to sell airtime, setting a precedent that would reverberate throughout the broadcasting world.

Hand in hand with the development of entertainment broadcasting came the phonograph record, to create an industry that became the largest of its kind in the world, with such brand names as Columbia, Victor and Brunswick. At first, the broadcasting stations never paid the artists for their services (except for the odd bouquet of flowers); there was no need. An invitation to do a solo performance was reward in itself.

As entertainment broadcasting began in the USA, it was developing on similar lines, but at a somewhat slower pace, in the various countries of Europe. As in America, the electrical and wireless manufacturers first began to take an interest in broadcasting. One of the first countries to have regular broadcasting was the Netherlands: the station was PCGG, owned and operated by the Nederlandsche Radio Industrie. Station PCGG at Eindhoven began broadcasting first for ships in the North Sea, but it very quickly acquired a much larger listening audience from as far away as England and Scotland, as its location provided an all-sea path to England.

Equally famous was the high-power telephony transmitter in England, built by the Marconi company for experiments with commercial wireless telephony, rather than entertainment broadcasting. Like its opposite number in Eindhoven it quickly acquired a following, made up from ships in the North Sea and experimenters, a band whose numbers had been swollen by those who had served in the Great War as signallers and wireless mechanics.

At this time the press began to take an interest in the activities of the experimental radio stations; after all, the subject was topical and it helped to sell newspapers. Radio broadcasting was seen as something of a novelty, associated more with entertainment rather than as a medium for the dissemination of news, and was therefore not a direct competitor with the press.

Typical of the reaction from the British press was that of the *Daily Mail*. With an eye for a publicity coup, it organised a concert to be broadcast from Chelmsford. Dame Nellie Melba, the Australian opera singer, was the sponsored solo artist, and on 15 June 1920 she travelled to the radio station in Chelmsford to make the first ever entertainment broadcast from Britain. Everything went off satisfactorily, and although the listening audience may not have been in large numbers, the broadcast was received by ships as far distant as St John's, Newfoundland.

Because few people owned radios, few heard the broadcast; the mass-produced set had not yet made its debut, and those who had receivers were mostly drawn from the band of experimentalists. Nevertheless, the press in general (and the *Daily Mail* in particular) made the most of the occasion, and entertainment broadcasting in Britain had passed its first test with flying colours.

Further broadcasts—mostly of an impromptu nature—took place from the Marconi factory at Chelmsford. It was typical of that era that those who operated the station were those who had designed it, and they became the artists, frequently making up the programmes as they went along. A few months went by, then, without warning, the government ordered Chelmsford to close down.

Thus radio broadcasting in Britain was suddenly stopped at the very time when it had demonstrated its popularity. The official reason for the enforced shut-down was that Chelmsford was creating chaos on the airwaves and interfering with military communications. In this sense, 1920 marks a milestone in British broadcasting: this was the first sign of an awareness by governments to the potential of broadcasting. With no precedent or previous experience to fall back on, it is likely that the government sensed potential in this new media but could not define it.

This abrupt closure of Britain's first radio station brought a sharp reaction from the amateur experimenters. These scattered enthusiasts had a voice, in the shape of the Wireless Society of London, formed as early as 1913 from an amalgamation of provincial wireless clubs. Dr J. Erskine Murray, president of the society, arranged for an official protest to be delivered to the Postmaster General requesting him to permit radio broadcasting to be resumed. This appeal was successful, and it was announced that Chelmsford should be permitted to resume its transmissions.

Looking back, it seems strange that a body of amateurs had the power and influence needed to make the government of the day change its mind. The most likely explanation lies in the First World War. Radio was still an infant, yet those enthusiasts who had served in the forces had used their skills to great effect. Radio had proved itself as a weapon of war: if war came again, these trained men would be invaluable to the nation.

The Marconi company's licence to resume its experimental broadcasts was subject to some restrictions; the radiated power should not exceed 250 W, transmission periods were limited to three minutes spaced with three-minute intervals, during which the station engineers were required to maintain a listening watch on the same wavelength. The purpose of this listening watch was to allow government stations to issue instructions to close down, if Chelmsford was found to be causing interference to military radio stations. The reduction of the radiated power to 250 W contrasts with the tens and even hundreds of kilowatts used by government commercial radio stations, which makes it unlikely that it was interfering with government communications. A more plausible explanation is that the government equated transmitter power with influence. This reasoning was to play an important part in the next decade.

The Marconi company resumed its experimental broadcasting, still under the direction of P.P. Eckersley—who performed as station engineer, announcer, programme director and even solo artist. The call-sign 2MT was allocated. Shortly after, the company obtained a second licence, 2LO; this station, located on the roof of a department store in London, was limited to a power of 100 W and came on air in May 1922. Government restrictions on these stations went far beyond mere technical conditions. Programmes were not allowed to be published, although the company was permitted to send out advance notice of its programmes to those on the company's mailing lists.

From the opening of the second radio station, events moved quickly, following the style and pace that had been set in America. Some 23 different manufacturers sought permission to carry out experimental broadcasting. Fearing chaos on the airwaves, the government proposed that these manufacturers, in conjunction with Marconi, should form a single broadcasting company. This was eventually composed of Marconi, Metropolitan Vickers, Western Electric, British Thomson Houston and the General Electric Companies. These companies carved up the country geographically, each building a radio station at the location of its factory. Stations were quickly installed at Birmingham, Manchester, Newcastle, Cardiff, Glasgow, Aberdeen, Bournemouth and Sheffield.

Although the British Broadcasting Company did not carry commercial broadcasting in the generally accepted way (sponsored programmes), the underlying basis of its operations was on a commercial footing. Each manufacturer was permitted to sell its own wireless receivers and receive a sales tariff on numbers of sets sold. In addition, each member of the broadcasting company received its share of 50% of the revenue received by the government on the issue of receiving licences.

In continental Europe, in places, a slightly different pattern was emerging. From 1919 to 1929 the radio station Konigs Wusterhausen, located about 30 km south-east of Berlin, transmitted press, economic and stock exchange reports in Morse Code. In December 1920 it broadcast several musical programmes, which were received in other countries in Europe. In 1922 the business broadcast on 4000 m was added.

In addition, some private firms transmitted test programmes—such as the C Lorenz AG on 1950 m and 3400 m alternately, from a site at Eberswalde, about 30 km north-west of Berlin. Some broadcasts took place from the house of the entrepreneur Dr S. Loewe directly from Berlin.

On 24 October 1923 Germany invented state broadcasting: the Reichpost-minister issued a 'decree on the introduction of an entertainment broadcasting service in Germany'. The required hardware was in position: an ancient submarine-type transmitter in the attic of the Vox-Haus in Berlin's Potzdammer Str 4. Its power was 250 W.

On 29 October 1923, at exactly 20.00 hours, a male voice was heard on the ether: 'Achtung, Achtung, Achtung—This is Station Berlin at the Vox-Haus on wave 400 metres'.

Chapter 3

Technological revolution

3.1 Power on the long waves

Marconi's achievement in bridging the Atlantic in 1901 was to produce a dramatic change of direction in the way telegraph communications around the world were being achieved. As early as 1902, stock exchanges in New York and London reacted swiftly. Shares in the cable companies slumped, with the realisation that wireless telegraphy threatened the existence of the all-powerful cable companies.

From 1902 the Marconi company was operating an experimental service across the Atlantic. In the same year Marconi received congratulations from King Edward VII and US President Theodore Roosevelt. By 1903 everyone was convinced that the days of the cable telegraph were numbered. Yet there was much to be done in the furthering of the art and science of long-distance wireless telegraph communications. In 1905 a major step forward was achieved when the Marconi company achieved a two-way communication, in full daylight, on a wavelength of 3660 m.

Thus the combination of high power and the use of long waves was seen as the future path of long-distance communications. This belief was to hold sway for another two decades, until the realisation of the existence of the short waves and exploratory work done by amateurs on them showed that the combination of high power and long waves had been a step in the wrong direction. But in the early 1900s no one could foresee this, and with yet further hindsight a technological gain resulted: the evolution of the first generation of super transmitters.

The first spark transmitter, constructed in 1901 and installed at Poldhu, used a 32 hp (28 kw) engine to drive a 25 kVA alternator. This produced 2 kV at 50 Hz, which was then stepped up to 20 kV and discharged across the spark gap. The secondary winding of the transformer was connected to the aerial and to earth. The primitive spark transmitter had many disadvantages, although at that time they were of little significance; chiefly there was the very poor frequency stability, coupled with a wideband occupancy and the resultant generation of countless harmonics from the damped spark.

By 1906 the state of the art had advanced. The spark transmitter had given way to the arc transmitter. Instead of alternating current direct current sources

Figure 3.1 *Marconi Wireless Station, Cape Cod, Massachusetts, USA, 1901*

Figure 3.2 *Marconi Wireless Station, Glace Bay, Newfoundland, Canada, 1901*

were used, with very high voltages, of the order of 12 kV, used to sustain the arc. Arc transmitters ran continuously, unlike the spark method, which was keyed: separate frequencies were used for the mark and space, thus eliminating the train of harmonics that appeared from the quenched-spark method. Transmitters using the arc method were built with powers of up to 300 kW. With arc transmitters, the key changed an inductance in the output circuit, thereby changing the frequency between mark and space conditions.

Although the arc transmitter was evolved from the spark transmitter with the object of eliminating the main disadvantages, it introduced yet another problem: the necessity of having to use separate frequencies for mark and space meant that the transmitter occupied a wide bandwidth. Nevertheless, it was an improvement over the spark method, and transmitters using the arc principle were built by the Marconi Company and Telefunken with output powers of 300 kW by 1906. Arc transmitters were operated from a high-voltage DC source, often from a bank of cells fed by a high-power DC generator. Possibly the first example of this method is that of a Marconi design, built for a transatlantic station at Clifden in Ireland.

The next stage of technological evolution was the high-frequency alternator, developed by Alexanderson in the USA. Applying rotary methods of power generation more usually associated with low frequencies, the high-frequency alternator was a major advance on any existing methods used in telegraph transmitters. It did away with crude spark methods and represented a major improvement over the arc transmitter. A typical high-frequency alternator example had an output frequency of 27 200 Hz and an output power of 200 kW. The shaft speed was 2675 rev/min and the alternator had 22 pole windings. The alternator achieved a frequency stability of 0.1% with the use of elaborate primary-compensation saturation transformers operating the frequency-control mechanism. Keying was usually by means of a magnetic modulator, which permitted telegraph keying speeds of up to 120 words per minute.

The high-frequency alternator became established as the most efficient method of generating a radio wave for telegraph transmission, and by the mid 1920s practically all of the major wireless stations had been converted to it, replacing the arc transmitter. Even so, this process took many years to accomplish because of the high capital costs involved: the economic life of a high-power transmitting station is of the order of 25–30 years, so there has to be a financial restraint on replacing equipments merely for the sake of keeping up with technology. The Post Office Radio Station at Leafield in the 1920s was a case in point. Leafield operated a high-power Poulsen-type arc for several years after most stations had abandoned them, and as a consequence it earned a reputation for the emission of harmonics and spurious signals.

The years between 1906 and the late 1920s witnessed the spectacular growth rate of wireless telegraph circuits on a global basis, removing any doubts on the importance of wireless telegraphy. The strategic importance of wireless communications in wartime had already been demonstrated: unlike cable, wireless communications cannot be severed. The use of long waves in conjunction with very high power had proved the ability of wireless telegraphy to span oceans and continents. The long waves possessed some advantages over the medium waves that had been tried: the long waves did not suffer from magnetic disturbances and gave a reliable service and consistent signal strength with no adverse signal variations of a seasonal or diurnal nature. In terms of usable distances, the high-power long-wave transmitter could provide a good service up to 8000 km (5000 miles).

By the 1920s the supremacy of the Marconi Company had been eroded by the challenges from its competitors, most notably the Telefunken company of Berlin. No longer did Marconi enjoy a monopoly on wireless communications. The number of countries operating long-distance wireless telegraph circuit had risen to 24. It is fair to point out that many of the wireless routes in the developing countries had either been installed by Marconi, Telefunken or the American companies RCA, Western Electric and AT&T. The state of the art in transmitter technology at high power was still the high-frequency alternator, and super power had begun to emerge. Between 1915 and 1925 the Telefunken company built several single and even dual 400 kW-transmitter installations around the world, at sites including Kootwijk, Malabar, Monte Grande, Torre Nova and Prado del Rey.

During this period Marconi had also been busy implementing its own wireless installations at home (Brentwood, Ongar and Caernarfon) and in many countries abroad. In 1919 the British government had decided that communication links with the British Empire should not be controlled by private enterprise; instead, the engineering department of the Post Office was vested with this responsibility, acting in conjunction with the appropriate authorities in the various Commonwealth countries. Reaction from the Commonwealth was not wholly enthusiastic: Australia and South Africa both objected, opting to retain full control of their own wireless stations. Marconi won the contracts to build these stations, both of the high-power variety: indeed, the transmitter station for Amalgamated Wireless Australia (AWA) was the highest powered transmitter ever contemplated: 1000 kW power, with a giant antenna system supported on 20 steel masts, 800 ft (240 m) high.

Super power had emerged around the world, but all high-power transmitters were still using the high-frequency-alternator or the Poulsen arc method. In 1926 a new star appeared at Rugby in England. This transmitter station, conceived by the British government as the heart of the Imperial Wireless Scheme, was unique in its design. Rugby Radio had been conceived in 1923 to take advantage of the advances made in tube technology. Its output power was comparable to, and even greater than, many existing transmitter stations; but its uniqueness lay in the fact that this was the first departure from the Alexanderson alternator. Rugby Radio, with the call sign GBR, came on air on 1 January 1926, making history as the world's most powerful radio transmitter using thermionic tubes. It earned another place in the history of radio when it provided a 24 hour daily service for 365 days a year, with a signal that could be received anywhere, at any time, night or day. This extraordinary capability was a result of the transmitter output power (350 kW), its antenna system, the excellent ground conductivity around Hillmorton, Rugby, and the fact that it happens to be almost in the middle of England.

The antenna system of GBR is a massive array of 12 masts, each 820 ft (250 m) high, supporting a long wave antenna with a circumference of nearly 24 miles (38 km). The GBR transmitter is still functioning today, though for an entirely different reason to that for which it was conceived.

Nothing lasts forever and the employment of high-power long wave transmitters for the purpose of carrying commercial wireless telegraphy was no exception. The demise of this mode of commercial communications began with the discovery and opening up of the short waves. Within another decade, commercial telegraph companies and government agencies had begun the migration to the short waves, which appeared to offer better prospects, enabling lower power transmitters to be used in conjunction with smaller masts and antenna systems.

Paradoxically, the cable telegraphy, whose demise had seemed so clear two decades earlier in 1906, outlived long-wave wireless telegraphy. This was attributable to two factors: the huge costs associated with high power, long wave installations, and the increase in commercial traffic from 1920 onwards, which gave a larger market.

Britain was linked to New York with a new telephone cable in the 1950s, to supplement those services carried by short wave transmitters from Rugby. One of the penalties imposed on Germany after the First World War was the loss of cable communications, and certain transatlantic cables were diverted to France and Great Britain. This was an indication of the strategic and political importance attached by governments to communications. In 1946 history was to repeat itself: Germany was deprived of its allotted frequency assignments in the long and medium wavebands. This step removed all possibility of Germany starting up international broadcasting.

Although high-power long wave telegraph transmitters are associated more with commercial than political roles there is at least one exception. Rugby Radio GBR has broadcast Reuters news service on 16 kHz, three times a day, since 1938. This service in plain language could be copied and used by anyone free of charge.

The leading press agencies had an untarnished reputation for reporting the truth from the world, no matter where that news came from. Even truth had its price, however. In September 1938 after Hitler's tanks had rolled into Czechoslovakia, the British government entered into a secret arrangement with Reuters. The Reuters news service would be broadcast from Rugby with an insertion written by the Foreign Office. The secret agreement provided that both Leafield Radio and Rugby Radio would carry 720 000 words per year, at a cost of three and a half pence per word.

During and after the Second World War, these two radio stations transmitted news whose content had been falsified with the intention of deceiving the enemy. There is a certain irony in the fact that the founder of the Reuters news service was the German Baron Julius Reuter, in 1851, when news was carried by pigeons. The Reuters experience is a sobering example of the extent to which governments will go to manipulate truth; it is also an example that political broadcasting ranks high in the armoury of a nation.

3.2 Discovery of the short waves

Although commercial broadcasters continued to use long and very long wavelengths until the late 1930s, the existence of another mode of propagation by short waves was known and under investigation from the early 1920s. The fall from favour of the long waves was not a sudden thing; despite the need for large amounts of transmitted power, the long waves demonstrated a stability that was not present at short waves. However, from the few experiments that had been conducted with the short waves certain advantages had come to light that merited deeper study.

These were to do with power and distance; the short waves seemed to have the ability on some occasions to travel very long distances with a minimal amount of power.

From about 1918 onwards, the mysteries of the short waves were being probed by a number of experimenters in America and Europe. This nucleus of workers was the forerunner of a worldwide fraternity of radio amateurs: a body of engineers, scientists and other research workers, but by far the greatest proportion made up of amateur enthusiasts. Mass production of radio receivers was still a long way off, and the only receivers in use were home-built from bits and pieces or experimental kits of parts.

Legislative control over experiments in wireless varied from country to country. In Britain, for instance, it was at first necessary to obtain a licence from the government even to construct receivers, and most countries exercised strict control over experimental broadcasting. The USA was a free society in this respect, with minimal government controls. As a result, its amateur radio movement flourished, and the contribution of American radio amateurs to scientific understanding of the short waves and the behaviour of the ionosphere was truly tremendous.

When long waves and super-power technology represented the state of the art, radio amateurs were assigned wavelengths below this band, as short as 200 m—considered of little use for serious communications. One reason for the banishment of radio amateurs to the short waves was the claim that radio amateurs were often the cause of serious interference to the commercial radio stations; in fact, much of this interference emanated from the broadcasting stations themselves as a result of radiation of harmonics.

Radio amateurs soon discovered that the 'short waves' of 200 m were far from useless. The wilderness to which they had been banished began to bear fruit. Communication was achieved between amateurs in America and France, with very small amounts of radiated power—watts, compared with hundreds of kilowatts being used by the LW broadcasters. But there were some baffling inconsistencies: for example, short distance communication was often impossible. The short wavelengths exhibited some other unsatisfactory characteristics: fading, unreliability of communications, and at certain times of the day and night, a complete loss of signal. This general behaviour pattern seemed to confirm what the commercial operators had suspected; that the short waves were too unreliable for commercial use.

The large body of radio amateurs in the USA, augmented by those in Europe, resulted in a co-operative movement that benefited by a cross-fertilisation of ideas and thoughts motivated to exploit the wavelengths below 200 m. From the apparently random experiences of the first experimenters, a more definite picture was emerging. Amateurs soon learnt that certain wavelengths, at certain times of the day and night, could give a predictable quality of service. Generally, it seemed that the lower the wavelength, the greater the distance over which communications became possible.

By the early 1920s international co-operation between the amateur radio societies around the world was becoming stronger. The American Radio Relay League (ARRL) worked closely with its counterparts in a number of countries in Europe, particularly those of France, Britain, Poland, Germany, Czechoslovakia and Hungary. In December 1921, the ARRL sent one of its experts, Paul Godley, to Europe with the best receiver available. Tests were run and 30 American

WIRELESS TELEGRAPHY ACTS, 1904–1926

LICENCE TO ESTABLISH WIRELESS TELEGRAPH STATION
FOR
EXPERIMENTS IN WIRELESS TELEGRAPHY

Mr. F. Wood and his son, James Wood as his agent...
of "Ivanhoe," 21 Minney Moor Lane, Conisborough, Doncaster, Yorks:
hereinafter called "the Licensee" is hereby authorized to establish a wireless telegraph sending and receiving station for experimental purposes at the above address......................
....
G3VG
....
....
....
subject to the conditions overleaf and to the payment of a fee of 30/- on the grant hereof (the receipt of which the Postmaster General hereby acknowledges) and a fee of £1 on the anniversary of the date hereof in each year.

This licence is subject to withdrawal or modification at any time, either by specific notice in writing sent to the Licensee by post at the address shown above, or by means of a general notice in the *London Gazette* addressed to all holders of licences for experimental wireless telegraph transmitting stations.

Failure to send the call signal or to tune accurately to authorized frequencies, the use of unauthorized power or frequencies, or any other breach of the conditions or non-payment of fees will render it necessary for this licence to be cancelled. In event of cancellation no part of any fee paid in respect of the current year will be returned.

This licence replaces that, dated the 27th October, 1937 which, with the call sign ~~which~~ is hereby withdrawn and should be returned to the address given below. 2AFR

Issued on behalf of the Postmaster General: C W Howe
25th October 1938

All communications should be addressed to The Engineer-in-Chief, Radio Branch, General Post Office, London, E.C.1, quoting Reference W2./A...5094...........

N.B.—Any change of address should be notified immediately

E-in-C 429 — 84551/37 — **(Prior authority must be obtained before the apparatus is removed or installed at a new address)**

Figure 3.3 *Transmitting licence for G3VG, Wireless Telegraphy Act 1904–1926*

amateur stations were logged. A year later the same tests logged 315 American stations. The next target was to achieve a two-way communication across the Atlantic, using low power and 200 m waves. Success came in 1923 when two American amateur stations contacted two French stations for several hours. A month later, the first communication was established with a British station. In 1924 the amateur movement achieved its most spectacular success when a British radio amateur, C.W. Goyder, a 16 year old schoolboy, established a two-way contact with New Zealand.

News of these most remarkable achievements sent shock waves through the commercial LW users. Large broadcasters began the migration to lower wavelengths: 200 m became assigned to commercial users and the amateur radio movement was forced to move downwards in wavelength to 100 m. When the 100 m band was found to exhibit similar characteristics to 200 m, the commercial radio stations were assigned this waveband.

With chaos threatening the frequency spectrum, an international conference decided that amateurs should be given specific wavebands. As a result, 80 m became assigned to radio amateurs worldwide, and this ruling was later amended to include 40 and 20 m.

In America, one of the first commercial users to become interested in the short waves was Frank Conrad, designer, constructor and operator of KDKA, the experimental station started by him, and the first regular broadcasting station in America. KDKA began experimenting, and made history for the second time by being the first radio station in America to broadcast on short waves.

In 1924 Conrad attended an international conference in London, which was also attended by David Sarnoff. Conrad had with him a single-tube radio receiver designed for SW reception. He invited Sarnoff to his hotel room, and using the hotel's curtain rod as an antenna he let Sarnoff hear the baseball scores from Philadelphia. Sarnoff was convinced that the future of radio was in the short waves. Such moments spelled the eventual doom of the giant LW Alexanderson alternator high-frequency generators, which had hurled their signals around the world with overwhelming power, and of the giant tube-equipped LW transmitters that were coming on stream as eventual replacements for the HF alternators.

By this time, the pioneering work of the amateur radio movement had stimulated some scientific investigations into the behaviour of short waves. The theory of Heaviside—that a bending of the radio wave was taking place above the earth—seemed to be the only possible explanation. The results of a number of workers (including Appleton, Breit, Tuve and others) into the physics of the upper atmosphere supported this. Even so, the behaviour of the ionised reflecting layers above the earth proved to be so complex that it took several more decades of research to be able to predict it with reasonable accuracy.

Understanding the cause enabled engineers to take steps to remedy the effects, using devices such as better designs of transmitting antennas to radiate energy in a low-angle directive manner and higher-gain antennas. Similar designs of antennas came to be adopted at the receiving sites; if anything, the receiving systems were more complex and costly than the measures adopted at the transmitting sites. Indeed, by the late 1930s the state of the art in high grade SW communication systems had reached the stage where vast amounts of money were being poured into practical measures to combat the seemingly random behaviour of the upper atmosphere. One such measure was the Multiple Unit Steerable Array (MUSA) developed at Bell Laboratories, USA—the home of many inventions in radio and communication sciences. MUSA was a desperate attempt to turn into a pure science what was then half art. It became operational in late 1940, designed to facilitate reliable means of telephony between London and Washington enabling Roosevelt and Churchill to conduct the war in Europe.

Another development of the late 1920s was the emergence of the SW relay. The discovery and the opening up of the short waves was like an Aladdin's cave. It made available the means whereby a broadcast on medium waves in mid-western America, destined for national US consumption, could be relayed across the

Atlantic, picked up at receiving stations in England, and re-broadcast over the BBC home network.

One of the essential prerequisites for relaying domestic programmes from one country to another was a good radio circuit over the SW link. This often involved much time being spent by the engineers at each end of the SW link carrying out a line-up test, and in order to do this, it was necessary to have good two-way communication in the first place. Sometimes this two-way communication was lost.

It was the private broadcasting companies of America—NBC, CBS, and others—that pioneered the use of short waves as a medium for entertainment broadcasting. In Britain and France, meanwhile, the medium and long waves were favoured. There were two reasons for this difference. First, the USA is a large land mass, which allowed short waves to be used for domestic broadcasting. Secondly, the USA had many commercial interests in Central America, particularly the Panama region. From the outset, US SW broadcasting was driven by advertising. Sponsors such as the United Fruit Company and Coca Cola had vested interests in reaching a wider listening audience than just the USA. The Federal Communications Commission issued licences to six companies for SW broadcasting. One of these was Westinghouse Electric, whose radio station W8XK became as famous as its counterpart on the medium waves, Frank Conrad's KDKA.

However, the development of SW broadcasting was not restricted to America: it was also taking place in Europe. But there was an essential difference of concept. Whereas the broadcasting companies of America perceived the short waves as another medium for domestic broadcasting, just like the medium waves, in Europe they were seen as a medium for international broadcasting, and as a means of maintaining contact with the distant colonies of Britain, France, and the Netherlands.

Chapter 4

Commercial broadcasting

The 1920s and 30s saw the very rapid growth of radio broadcasting in North America and Australia. In contrast to Europe, where growth in broadcasting was proceeding in an orderly but not too spectacular fashion, in America there was a hectic scramble to get on the air by whatever means. The difference lay in the regulation of radio broadcasting: in most European countries, it was perceived as a mass medium that needed to be under the control of the state. In Britain, for example, legislative control over radio broadcasting was total. It extended beyond the technical constraints and included a rigid control, amounting to censorship, of the social and political aspects. There were no such restraints in America, and government exercised the minimum of technical control. All broadcasting was owned and operated by private capital.

KDKA's broadcast in 1920 of the Harding-Cox presidential election is usually cited as the historic beginning of regular radio broadcasting in America. Nevertheless, a number of other stations in the USA could claim the honour. For example, KQW in San Jose, California, first transmitted a radio programme in 1909, and was running a regular schedule by 1912. A Detroit amateur station, W8MK, began regular transmissions a couple of months before KDKA. From the historical viewpoint, however, KDKA meets the criteria that guarantee its place in American broadcasting history: it was the first to begin broadcasting on a daily basis, it broadcast organised programmes for the general public, rather than for experimentalists, and it was the first radio station to be licenced by the FCC as a broadcasting station. But it was undoubtedly the broadcast of the results of the Harding-Cox election that gave KDKA the edge over its competitors.

KDKA's owners Westinghouse did not have its monopoly for long. Owning, or having access to, a radio station began to have a strong appeal to major department stores, newspapers, educational institutions, churches and even electrical-supply merchants, but most of all, to the giant electrical, telephone and telegraph companies such as General Electric, Western Electric and AT&T. The motivating force behind the owning of a radio station was not always the same. To some it was a means of propagating the gospel of God, to others it was a way of selling services and goods. To the giant electrical companies it was also a status symbol, promoting the name of the company throughout America—and a tax-deductable company expense.

By the end of 1920 some 30 radio stations had been licenced. In the spring of 1922, a new industry – the manufacture of radio receivers – was taking shape, and by the end of that year 100 000 radio receivers had been sold (no licence was needed to own a receiver). By May 1922, some 200 broadcasting licences had been issued in the USA and a year later the number had risen to 576. While radio broadcasting in Britain was still locked in a straitjacket of bureaucracy and stifled in an atmosphere of social snobbery, broadcasting in America was free for all to enjoy and be dazzled by the sheer variety of programme content and style.

In the beginning, US broadcasting was not advertising-driven. Broadcasters had a variety of reasons for owning and operating a radio station; some wanted to further a political career, or fulfil a personal ambition. Radio station WEAF (flagship of the telephone group AT&T) and Western Electric began to sell air time in 1922. Thus WEAF planted the seeds of a new business that eventually grew to envelop the broadcasting industry: advertising, public relations and propaganda. From about 1927 the revolution was under way. Advertising agencies, manufacturers, sponsors, promoters, and the sellers of medical and life insurance were jockeying for places in a world of propaganda disseminated by radio broadcasting. Propaganda comes in many guises, each with its own objective: social, political, religious, this one was to do with advertising. The best that can be said for this kind of propaganda is that it does not kill people.

The driving force in this new revolution was the PR industry. Advertising executives worked closely with the sellers of air-time (broadcasting companies) and the buyers (product manufacturers). Some of the most powerful and influential lobbies in America were the big industries: tobacco, automobile manufacturing, drugs, and fruit (which itself owned ships and radio stations). It was inevitable that such a powerful group would exert considerable influence over the radio broadcasting industry. Script writers, producers and directors all came under their influence in one way or another. Through the medium of sound broadcasting, brand names such as Kleenex, Pepsodent and Lucky Strikes became household names creating a volume market of demand from the great American public. Three weeks after Kleenex (the first paper tissue) was launched on the market, an advertising man exclaimed: 'Its going great – women are actually throwing them away!'

By the 1930s the numbers of radio stations in America had grown to nearly 1000. The sheer size and area of the land mass made it possible for radio stations that were separated geographically by hundreds of miles to operate on the same assigned frequency. The size of America also encouraged the use of the short waves. It was easier to broadcast to Panama with a 20 kW SW transmitter in Boston than to set up a MW transmitter. Thus, the USA gained a lot of experience and expertise in SW broadcasting ahead of many European countries.

Competition usually enhances the quality of a product, and American broadcasting was no exception to this rule. Inevitably there were many casualties along the long and rocky road to excellence, but eventually America gained half-a-dozen companies worthy of some note: Columbia Broadcasting System (CBS), Crosley Radio Company (CRC), General Electric (GE), National Broadcasting Company (NBC), Westinghouse Electric (WE), and Worldwide Broadcasting Company (WBC).

The major sponsors could afford to buy air-time on the best broadcasting networks, and to buy the best artists to be found in the USA. From this evolved

Figure 4.1 *OSL cards of some famous shortwave stations of the 1930s: W8XK, VK2ME, W2XAF*

a pattern of sponsors creating their own radio shows, with such shows as the Kellogg Hour. The 1930s was also the era of the big bands: Paul Whiteman, Count Basie, Duke Ellington, Cab Calloway and many others. As a result, listeners in America could hear radio shows that listeners in Europe could only dream about. The range of diversity and quality contrasted with the BBC's staple fare of chamber music, string quartets, opera and after-dinner speeches, which had little appeal to the ordinary working man.

Excellence was not confined to the quality and diversity of programmes: America was also building a mammoth manufacturing industry. Big companies such as GE, Westinghouse and RCA owned many different brand names in radio receivers. These same companies invested huge sums into developing other broadcast equipments for professional users, and the experience they gained in broadcasting was put to good use in developing some outstanding transmitter designs. By the mid-1930s, America had half-a-dozen manufacturers of transmitters with such features as high-level modulation, and output powers of 20 kW or so. The East Coast of America, particularly around New York, New Jersey, Pittsburgh and Boston, was the centre of the broadcasting industry.

With the growth of SW stations in America, which carried the same shows as the MW stations, listeners in Europe regularly tuned into stations such as W2XAF and W2XAD Schenectady, owned by GE, W2XE and W2XDV Atlantic City, owned by Atlantic Broadcasting Corporation, NBC's W3XAL and Westinghouse's W8XK. Eager listeners in Europe soon discovered that the reception of these East Coast stations from America was not confined to the short waves. During the winter evenings, from 10 pm onwards, it was possible to hear the East Coast stations broadcasting on medium waves (when an all-sea, all-darkness path gave the best possible ionospheric reflection).

Canada, although part of the British Empire, wisely did not follow the example of the mother country by placing broadcasting under the control of body working to a royal charter. Canadian industries owned and operated commercial radio stations on the same lines as their US counterparts. At that time wheat growing was the biggest industry in Canada, and one wheat-growing company, James

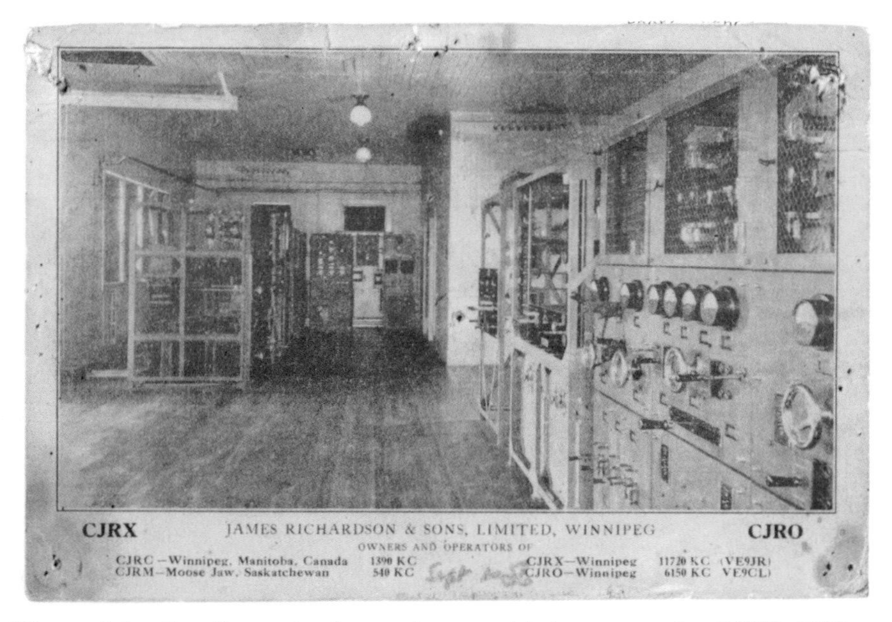

Figure 4.2 *Canadian, privately owned commercial shortwave station CJRX, 1936*

Richardson & Sons Ltd of Winnipeg, owned and operated six SW and MW radio stations. I heard Richardson's SW station CJRX on 22 September 1936, on a frequency of 11730 kHz. At the time the station had a power of only 2 kW, which was very low for a broadcasting station. The state of the technology is evident from the contemporary picture of the front panel of the transmitter.

Australia in the 1930s had few radio stations, but being the most distant of Britain's colonies it had an awareness to the importance of high power. Amalgamated Wireless Australia was the government-owned broadcasting service with responsibility for all external communications and broadcasting. In Britain, the most famous radio station operated by AWA was VK2ME, Sydney, which could be heard in England at about 6 am, with some patience. It was easily identified by the mocking call of the Kookaburra bird, which it used to open and close the broadcasts. VK2ME had a SW carrier power of 20 kW.

Chapter 5

A social tool: birth of the BBC

The British Broadcasting Company was formed in 1922. Although the original intention may have been to disseminate entertainment to the masses, radio quickly became a social instrument, seen as the exclusive preserve of the upper-middle classes. It was quite normal for the average listener to be depicted as dressed immaculately in full evening dress, seated or standing elegantly with an expensive brand of cigarette in his hand, listening to his set.

The BBC was happy to live up to this stereotype. Radio announcers always arrived in evening dress, and announcers were chosen from the upper classes of English society. Most importantly, they had to be able to speak the King's English just as the King spoke it. By the end of 1922, nearly 50 000 receiving licences had been issued; this represents roughly one per 1000 population. Naturally, these were drawn from the wealthy upper-middle classes.

The British Broadcasting Company officially came into existence in November 1922, although its licence was not received until the following January. As well as the technical control that was exercised by the Postmaster General on behalf of the government, he retained control over broadcasting extending to the spoken word. Of course, not every word broadcast by the BBC was written down and checked beforehand, but the control was nevertheless tight. Any subject that could be considered contentious, political or likely to offend certain sections of the public could not be transmitted. Naturally, this ban included political talks, and particularly anything to do with socialism. One week in January 1923 included a surfeit of opera: Marriage of Figaro, Faust, Pagliacci, Siegfreid, Hansel and Gretel. By way of light relief, the first broadcast of dance music took place in May 1923, although even this was intended strictly for the upper classes. It was a broadcast by the resident orchestra from the Savoy Hotel.

Possibly the best indication that the government was aware that broadcasting had a potential—but one which it could not define—was the fact that initially no news was permitted to be broadcast. News remained strictly within the control of the press barons. As the years progressed, there was some variation in the type of programmes: there were talks for men and talks for women, and military music by the guards regiments was evidently considered worthy of broadcasting on patriotic grounds.

The growth of radio broadcasting in Britain was stifled by the cost of the receiving licence and the cost of a radio receiver. During the 1920s, the average working man in Britain earned about £68 per year; a motor car cost around £450, and a radio receiver might cost anywhere between £25 and £100. Thus, it will be seen that the radio set required a purchasing power comparable to a car in Britain today.

This was reflected in the way radio receivers were marketed. The possession of a radio set in the early 1920s had the same importance to the lower middle classes as did the car 20 years ago. The possession of a radio set did not merely signify a social status; it was an important symbol of personal prestige to the working man. Like the car of today, the working man ideally sought to change his radio set each year, trading in his old set in part exchange. Changing sets for the latest model with more valves became an all-consuming passion for those who could afford it—and for those who could not, it was made easier by hire purchase.

Also like the car of today, brand names and models were graded in terms of numbers of valves: three-valve, four-valve and even up to ten-valve sets. Models were given names intended to signify power and performance. Over a period of five years my father acquired models of ascending power: the Ultra Cub, the Tiger, the Panther and finally the Super Panther.

From 1922 the BBC as a limited company broadcast entertainment programmes and social events to the people of Britain and Ireland. It did this with ten main transmitting stations and a similar number of relay stations. Technology at that time did not permit the construction of high-power transmitters or antenna systems. As a result, reception in many parts of the British Isles was far from satisfactory, and in certain areas it was non-existent. However, it should be remembered that the same was true of other countries, including large areas of Mid-West America.

The BBC developed its skills in transmitter engineering largely under the guiding genius of its first chief engineer, P.P. Eckersley. It was he who steered the development of its transmitter network, made possible by the related developments in the manufacture of water-cooled transmitting triodes.

In 1925 the future of the company came under government review, as its broadcasting licence was due to expire at the end of 1926. In just over four years, this private company had raised the technical standards of broadcasting in both studio and transmission technology. Moreover, the number of fee-paying licence holders had risen to over two million. This growth, along with similar developments in America and Europe, had demonstrated some of the possibilities and awesome potential behind this new science. It was now time to take this power out of private control and place it firmly under the control of the government.

In December 1926 the BBC as a company was dissolved, and replaced by a Corporation operating under a Royal Charter. Instead of a managing director, it now had a director general. This new post was filled by the old BBC's general manager: J.C. Reith.

Reith had joined the company in 1922, with no background experience of broadcasting. He had previously been an army officer, and later the general manager of an engineering works in Glasgow. Reith sometimes described himself as an engineer, and on other occasions as a civil engineer. Nevertheless, whether by accident or design, the government had found the right man. He ruled with a rod of iron from the first day of his appointment as general manager; when he became director general, he immediately received a knighthood, but his style of leadership did not change in the slightest.

The new BBC was placed under the control of a board of governors, 'persons of judgment and independence.' However, the terms of the Royal Charter leave no room for doubt that broadcasting was seen from the outset as an instrument for serving the interests of the government, rather than a means of providing an entertainment service. In the licence which came with the charter the Postmaster General retained authority to approve wavelengths, heights of masts and hours of transmission; he was further invested with powers to take over or close down the BBC in case of emergency.

As well as this technical control, the government also exercised political power. This was made evident during the General Strike of 1926, when some provisional arrangements were made to take over the BBC. The word used—'commandeer'—leaves little doubt that such a move would have been with military force if necessary, and on 15 November 1926 the Postmaster General told the House of Commons that he had instructed the BBC not to broadcast its own opinions on matters of public policy.

During the General Strike, the BBC's transmitting stations were guarded against an attempted takeover. The BBC was prohibited from making comment on the strike, or even reporting on the happenings in the country; news was compiled by the government for broadcasting, and on one occasion the BBC engineering

Figure 5.1 *Aerial system for 2LO BBC, 1927, on roof of Selfridges Department Store, London*

Table 5.1 *Milestones in British broadcasting, 1920–1933*

Date	Event
23 November 1920	Government imposes ban on broadcasting following transmissions by the Marconi company
18 October 1922	British Broadcasting Company formed by six manufacturers
15 December 1922	BBC registered as a limited company
23 December 1922	First orchestral concert broadcast
8 January 1923	First full opera broadcast from Covent Garden
21 January 1923	First symphony concert broadcast
23 January 1923	First military musical concert broadcast
24 January 1923	First broadcast of late-night dance music
9 October 1923	Broadcast of the Lord Mayor's banquet
26 November 1923	First experimental relay to and from the USA, with KDKA Pittsburgh
1 January 1926	BBC becomes a Corporation under a Royal Charter
4 January 1926	Compulsory for station announcers to wear evening dress
15 November 1926	Postmaster General reveals political ban placed on the BBC during the General Strike
1 November 1927	Experimental transmission by short wave to the Empire
19 December 1927	Empire Service is inaugurated by King George V
15 March 1933	Radio Luxembourg comes on air with 200 kW of power
24 March 1933	BBC accuses Radio Luxembourg of pirating
7 April 1933	BBC asks the Union Internationale Radiophone (UIR) to prevent Luxembourg from broadcasting (to England)

Between 1933 and 1937 the BBC continued to press through the Foreign Office to ban Radio Luxembourg from broadcasting to Britain.

department was instructed to overmodulate the transmitters, thus increasing the output power. This may be compared to the similar but much more sophisticated techniques devised in 1950 to counteract Soviet jamming by barrage broadcasting.

Although the General Strike lasted for eight days in May 1926 it was the long-running strike by miners in Britain's coalfields that posed the biggest threat to the government. In the end, the strike brought poverty and humiliation to whole communities of coal workers. It also had the effect of concentrating the mind of the government, as never before, on the need to exercise strong political control on the country. This also focused political attention even more strongly on the BBC. By the late 1920s, the BBC was reaching out to a large sector of the community, including many of the working class. However, there was little change to be seen in the types of programmes: there were a few comedy broadcasts, and dance music was now broadcast regularly, but this was a concession to the upper class in general.

Reith believed passionately that it was with God's will that he was in charge of British broadcasting, and that it was his destiny. He also believed that monopoly in the case of broadcasting was a virtue, and it gave him the duty to choose and broadcast the kind of programme he thought was good for the British public—rather than the kind of programmes the general public might have chosen for itself. Broadcasting standards were, if anything, tightened after the Royal Charter. The wearing of evening dress by the station announcers, although it had always been general practice, now became compulsory; this was one of Reith's first changes as director general.

From the point of view of the working man, radio programmes became even more dismal. British broadcasting was staffed and run by men who had not come from the working class and did not understand the working man, but presumed to understand what the working man wanted. Britain occupies a unique place in broadcasting history in that it was the first nation to see this new medium as an instrument of social manipulation rather than for entertainment. It was also the first country to perceive radio broadcasting as a means of dissemination of propaganda on a national level. The seeds had been sown for the perception of radio in an even more important role: as an instrument of foreign policy whose function would be to disseminate informational propaganda to the rest of the world.

Chapter 6

Propaganda: the cradle years, 1927–1938

From 1927 onwards, the short-wave radio transmitter became an instrument of foreign policy for the specific purpose of information broadcasting—more correctly termed propaganda. Great Britain was one of the first countries to realise that the short waves were the perfect medium for communicating with its far-flung colonies in the eastern and western hemispheres of the world.

In the same year, the Soviet Union and Germany began their SW service. By 1929 the USSR was broadcasting its programmes on the short waves in French, English and German. Germany too began to radiate its programmes in a number of different languages. Great Britain, of course, used only the one language—English—in all its broadcasts, presumably because the programmes were intended for the intelligentsia, and not for the foreigner in the street.

Satisfied with the results from its experimental series of transmissions, the BBC was given permission to embark upon the building of a more permanent SW transmitter complex at Daventry, England. The SW transmitter G5SW at the Marconi site at Chelmsford was relegated to history, to be replaced with a more modern installation. Initially the site at Daventry accommodated two transmitters, modulated at low level, with output powers of 10–15 kW.

Because of the global nature of the British Empire, a great deal of thought was given to the choice and arrangement of the antennas. It was decided to use directional antennas for each of five zones of the world, centred on Australia, India, South Africa, West Africa and Canada. Because of the vast size of these different territories, the antennas had to produce much broader beam widths than antennas used for point-to-point SW services.

Daventry SW station commenced testing by 14 November 1932, and on 25 December 1932 King George V became the first monarch to speak to his peoples throughout the world. The transmission was directed first to Australasia, and at intervals throughout the day the same message was repeated and beamed to other parts of the world. The first royal Christmas broadcast began with the words: 'Through the marvels of modern science, I am enabled this Christmas Day to speak to all my peoples throughout the world.'

Broadcasting to the British Empire was instituted as a regular service from 1932. It was designed with the purpose of keeping Britain's colonial civil servants in touch with the mother country; also, because many of the colonies were English

Figure 6.1 *Empire broadcasting on shortwaves. G5SW Daventry, England, 1927*

speaking, the empire service helped to keep British culture alive in the minds of the subjects of these colonies.

Reith had contemplated the idea of creating an empire broadcasting service as far back as 1924, but had not met with much enthusiasm on the part of the government. Strong support from the British Commonwealth Office for his scheme did not come until 1928–29, which date seems to mark an increased awareness to the use of the radio transmitter as an instrument of foreign policy. Subsequently Daventry was equipped with six short wave transmitters by 1932.

Techniques in station announcements and identification were developing. The BBC settled on broadcasting such identification signals as the voice of the nightingale bird and the strident chimes of Big Ben. 'London Calling' became the announcement that identified all broadcasts of the empire service. Although the British Empire was predominantly English-speaking, there were very large areas of the world where it would have been advantageous to broadcast in tongues other than English—Arabic, for example, would have been an obvious choice as a second language.

The mid 1930s was the turning point in an increased awareness to propaganda broadcasting. During this time, Italy became the first country in the world to broadcast in the Arabic language. From its state-of-the-art Prato Smerelda Radio Centre in Rome, the Italian government initiated Arabic broadcasting in 1934. Early experiments were not blatantly political. Radio Bari, as the service was known, faced many problems, such as lack of radio receivers, and programme timing difficulties in a region of the world where time has a different dimension to that in Europe. With some ingenuity and thought, the Italian government distributed some radio sets, and to overcome the problems of timing Radio Bari used known terms such as sunrise and sunset when announcing programme times.

One of the first examples of reprisals and government censorship in international broadcasting took place in Europe over the Ethiopian Crisis, when Benito Mussolini's army moved across the border into the highlands of East Africa. Much international attention became focused on this issue between Italy and Britain. Britain had strong vested interests in this part of Africa, and proposed sanctions against the Italian government, to be discussed by the League of Nations in Geneva.

The Columbia Broadcasting System (CBS) had the idea of covering the debates by relaying to America the cases presented by both sides. CBS was an exponent of the art of SW relay—a fairly primitive technique compared to modern satellite links; SW relaying worked well but was a laborious process which involved setting up a SW link and co-ordinating the activities at both ends.

Edward R. Murrow of CBS was due to cover the events. On 9 October 1935, with the war one week old, the chief Ethiopian delegate spoke over the Geneva–London–New York SW hook-up and presented the case for the government of Haile Selassie. All went well with the arrangements, the broadcast was received at Riverhead SW receiving station and re-transmitted over the domestic broadcasting network for Americans to hear.

The next night was the turn of the Italian delegation to present the case for the government of Italy. The SW link had been given a line-up test, Murrow waited in the chamber with the Italian delegate Baron Pompeo Aloisi. In New York the time was approaching 6.15 p.m.—time for the relay to come through. With a few seconds to go, the telephone in the Geneva studio rang: it was a message for Edward Murrow that the relay had been cancelled. The British Post Office, which controlled the British end of the SW link with New York, refused permission for the broadcast to go ahead and had pulled the plugs on the circuit. Murrow, through the engineers at Riverhead, had the unwelcome task of advising CBS in New York there would be no Italian broadcast that night, nor any other. CBS in Geneva sent off a query to London and the following day Murrow received his answer: 'In view of the imposition of Article XVI of the League of Nations Covenant, it was thought neither desirable nor proper to extend British facilities to the Italian delegate.'

On Friday 11 October the event was front-page news, with headlines: '51 nations vote for sanctions against Italy, Britain cuts off the broadcast by the Italians'.

International broadcasting and anti-broadcasting had arrived as a weapon of considerable effect for influencing the minds of listeners.

Eventually, the Italian government did get its broadcast from Geneva to New York through its own SW circuit, which was every bit as good as the London link.

6.1 Broadcasting to the Arab world

In contrast to the emphasis on the written word in western culture, Arab society was (and still is) heavily orally based. In these conditions, the radio programmes broadcast from Rome became a part of the way of life for the *fellahin*, who after their day's toil would betake themselves to the primitive cafés, under the fuming oil lamps, smoke their pipes and enjoy the communal loudspeaker entertainment.

From 1937 Radio Bari became more political in its nature, as Mussolini began taking a greater interest in Ethiopia and other Arab-speaking countries. Mussolini's interest was understandable; large parts of North Africa were Italian colonies, such as Cyrenaca (Libya). However, the Bari broadcasts gave the British government much concern. Britain had always regarded the Middle East as within its rightful sphere of influence. For many decades it had exerted considerable influence in Egypt, Palestine, Trans-Jordan, Iraq, Persia and the Persian Gulf. Now, for the first time, another European power was achieving a certain degree of success in wooing Arabic-speaking peoples.

1937 was also a testing time for Britain's authority at home and in other parts of the world, due to state events and happenings in the royal family—the abdication of King Edward VIII followed by the coronation of George VI. During this crisis the BBC used all its available senders to broadcast news of these events to the world; but with these events of state out of the way, the BBC was able to turn its attention to projecting Britain's influence to the Arab world, and countering the growing Italian influence.

The British government had already established several committees to determine how best to combat the effects of Italian propaganda broadcasting. In late 1937 the government authorised the BBC to start up an Arabic service in opposition to the Bari broadcasts. This service commenced on 3 January 1938, and the inaugural broadcast was made by the Emir Seif-al-Islam Hussein, son of the King of the Yemen.

On 29 October 1937, the Postmaster General announced that the BBC would undertake broadcasts in other languages. This was quickly followed by a clarification from the Chancellor of the Exchequer to say that the BBC had been 'invited' to make foreign-language broadcasts in Spanish and Portuguese to South America, in addition to Arabic broadcasting for the Middle East. According to some sources, there was evidence that the Spanish and Portuguese language broadcasts were authorised partly as a smokescreen, to avoid giving Mussolini the impression that the Arabic service was intended solely to counter the Bari broadcasts.

The BBC enjoyed much success with its Arabic broadcasts: problems of personnel, dialect and format were resolved so that it was able to set a standard of service and gain a reputation for being a reliable source of news and information, objectively reported and independent of government control or censorship. The propaganda war between Britain and Italy continued for several months, and was officially ended on 16 April 1938 with the signing of the Anglo-Italian Pact. This pact recognised for the first time the significance of radio propaganda as a diplomatic instrument and a political tool.

However, there is much evidence that Italy never forgave the British for their Arabic propaganda broadcasting and that this incident was one reason for Italy throwing in its lot with Nazi Germany shortly afterwards. In late 1938—about the time of the Munich agreement—Italy turned its propaganda efforts towards Europe, whilst its partner Germany took over the role of broadcasting anti-British propaganda to the Arab countries. Germany enjoyed much success in the Middle East with these broadcasts. One of its key announcers was Yunus Al-Bahri, whose microphone skills were said to be unmatched even into the 1950s and 60s in the Arab broadcasting world.

From the end of 1937, propaganda broadcasting by European nations had become a fact of life establishing its role as an instrument of 'cold war'. In Britain there were voices to the effect that the countering of foreign propaganda broadcasts was far too important to be left to the BBC. This was an opinion that would grow much stronger in the coming years.

1937 was also the year when the House of Commons passed a motion recognising the need to counter totalitarian propaganda broadcasting, not by retaliation, but by the widespread dissemination of information and news. This curious ability of the British establishment to draw a fine distinction between 'propaganda' and 'information', with enemies associated with the former and Britain with the latter, has been assiduously fostered over many decades, extending to the staff of the BBC itself as well as the general public; yet the two are one and the same. In Germany the opposite definition held sway; the true meaning of the word propaganda was upheld to the extent that the German media never used this term in reference to foreign broadcasts, but only in the context of information broadcasting by the Third Reich.

6.2 Nazi propaganda

When his National Socialist party came to power in the 1930s, Hitler recognised the importance of strategic communications, including road, rail and air transport, and radio broadcasting. SW communication centres and broadcasting complexes were built by distinguished companies such as Telefunken and Siemens. At Zeesen, 20 miles south-east of Berlin, a new state of the art SW transmitter centre was built by Telefunken. Soon, 12 SW transmitters were beaming the voice of Nazi Germany to all parts of the world.

Zeesen, using call signs extending from DJA to DJR, beamed its broadcasts on the 49, 31, 25 and 19 m bands. German broadcasting was refined and entertaining, with concerts, competitions and free Nazi-type material being sent out to listeners. German SW broadcasting was a positive force intended to promote German unity and its quality of life. The Nazi Swastika—the symbol of goodness—adorned the QSL cards sent out by its SW stations. Although the programmes broadcast in the mid 1930s at times criticised certain aspects of British and colonial rule, German propaganda reserved its most virulent attacks for its arch-enemy the Bolsheviks. Even during the period from the Munich crisis, Nazi broadcasts did not attack British imperialism to any great extent; Britain's

Figure 6.2 *Zeesen SW station of the Third Reich, a propaganda broadcaster, 1937*

achievement in building an empire was one to which Hitler aspired, and sought to emulate in his creation of an all-Europe empire. However, German propaganda broadcasts were anti-Jewish to the extreme, and this was interpreted by many in the British government as hostile to Britain.

6.3 Soviet propaganda

Last of the big four European propaganda broadcasting nations was the USSR. Soviet broadcasting started in 1922, about the same time as many other European countries. Its evolution was dictated to a large extent by the country's very large land mass, which extended almost half-way around the world: SW broadcasting was best suited in many respects to the very large countries in the world, and this is still the case today. Therefore the USSR took to SW broadcasting ahead of many other countries in Europe.

Radio Moscow, the external broadcasting service of the USSR, went on the air in 1929 carrying propaganda broadcasts in German, English and French. The word *small* is an adjective that plays no part in the context of the Soviet Union, and its broadcasting stations were no exception to this rule.

Soviet broadcasting from the mid 1930s concentrated its efforts in promoting the Communist way of life, telling its listeners about Soviet achievements with its five-year plans and extolling the efforts of the people. Although Communism and Nazism are opposite poles, there was a certain degree of commonality in the two styles of propaganda broadcasting: both were positive in character when

Figure 6.3 *Radio Moscow. An unusual QSL card from the shortwave station RNE Moscow. It commemorates the Soviet expedition to the pole*

describing the achievements of their country, and both nations reserved their virulent style of propaganda for the other. The Soviets expended much time and effort in warning the rest of Europe about the dangers of German National Socialism.

By 1938 propaganda broadcasting in Europe via the short waves had become established, and had given ample evidence of its potential as a weapon of cold war and in laying the seeds of a real war. However, the development of SW broadcasting in the Europe of the mid-1930s would be incomplete without reference to those countries that pioneered the art of SW broadcasting for other reasons. America was the first country in the world to use SW broadcasting for another kind of propaganda: the indoctrination of listeners into buying radios, automobiles and cigarettes.

Chapter 7

Radio Luxembourg: super power comes to Europe

By the mid 1920s, the mathematical relationship in LW telegraph communications had been established: the greater the power of the transmitter, the greater the distance the signals travelled. This same logic applied to sound broadcasting on long and medium waves, except that both area and distance counted, since broadcasting is usually omnidirectional in character. This relationship between power and area had been known for a long time, but the technology in the shape of the high-power, water-cooled triodes did not exist until the 1930s, thus preventing the construction of high-power transmitters for sound broadcasting.

The first super-power broadcasting station was Radio Luxembourg in Europe, closely followed by the emergence of an even more powerful broadcaster in the USA: WLW Cincinnati. Radio Luxembourg came on air in 1933, one year before WLW. By 1934 there were several broadcasting stations in Europe with output powers approaching and even exceeding 100 kW carrier power, but Radio Luxembourg generated a full 200 kW.

From the beginning of sound entertainment broadcasting in Europe, the long and medium waves became the accepted medium for broadcasting on a regional and national basis, while the short waves were the accepted medium for international broadcasting. With this curious perception it is not surprising that each country thought it had a monopoly on national broadcasting.

This reasoning on the part of many broadcasting authorities in Europe would explain the high-handed attitude of the BBC and Sir John Reith, who believed it was his duty to broadcast the type of programme he thought good for the people, rather than what the people would have chosen for themselves. At no time during the first decade of its history did the BBC try to find out the kind of programmes its listeners would have liked.

When the monopoly in national broadcasting by the likes of the BBC was broken by the ability of listeners to tune in to other stations on the Continent, it demolished the theory that high power would automatically result in a high listening audience. In the world of radio broadcasting (and this includes television), the most powerful of super transmitters is useless if it cannot command a size of audience commensurate with area served. This can only be achieved by methods of appeal, selecting the right programme for the listeners for whom the broadcasts are intended. From the first day Radio Luxembourg came on-air it showed that it had found the formula. It took Europe by storm with its popular style of programmes, transcending class barriers and national frontiers.

Such was the huge success enjoyed by Radio Luxembourg that within two years it had captured a 50% share of the listening audience throughout Europe. Even more remarkable was the fact that out of 4 million listeners to Radio Luxembourg, 2 million were in the British Isles. This was twice as many listeners as the BBC.

The high esteem that Radio Luxembourg enjoyed in Britain is best understood in the context of the social conditions of the mid 1930s. Although it was the mother country of the British Empire, a large sector of the population (15–30%) lived in conditions of chronic poverty. Those fortunate enough to be in employment worked a six-day week, Sunday being the only day of rest, but even this was overcast by the fact that it was necessary to retire to bed early for the Monday morning start.

Yet Sundays were the very days when the BBC chose to dole out the most dismal of programmes, a mixture of church services, classical music and organ music. The coming of Radio Luxembourg was like a breath of fresh air sweeping through the air waves of Europe. Its entertainment programmes were like nectar to the working man; they turned a dismal day into a pleasure to be looked forward to.

For technical reasons Radio Luxembourg shifted its operating wavelength no less than three times during 1933, and during these frequency changes the listeners voted with their fingers: where Radio Luxembourg went, the listeners followed. The station was truly the Pied Piper of broadcasting. After broadcasting on 1250 metres, 1185 metres and then 1191 metres, it finally settled on the wavelength of 1304 metres. Radio Luxembourg now had a permanency and a potential that bothered the BBC.

There were many voices – particularly in the BBC – of the opinion that Radio Luxembourg was a 'low-brow' type of radio station, and not therefore an example for the BBC. Yet the list of its sponsors seems to disprove this theory: its shows promoted products that included cruises and holidays abroad. The sort of companies that advertised on Radio Luxembourg ranged over a wide social spectrum, with food and medical products being much in evidence.

A further, important reason for the success enjoyed by Radio Luxembourg was the quality of its programmes, both in artistic merit and in technical excellence. Under the overall control of the Walter J. Thomson public relations and advertising company, its studios in Bush House produced up to 44 shows per week. Because there was no available land line from London to Luxembourg, it used the very latest state of the art in recording techniques. The recorded programmes were taken by road or air to the station's studio in Luxembourg.

The irony in the situation was that Radio Luxembourg was driven by British marketing expertise, Dutch studio technology (in the shape of the magnetic tape recording system) and French and German transmission technology, much of which was made by Telefunken in Berlin. The BBC had good reason to be envious of the success enjoyed by Radio Luxembourg.

The BBC, supported by the Postmaster General, strove to silence Radio Luxembourg, and even to force the station off the air, by whatever means came to hand. In this case it would seem that the end justified the means, because the BBC even considered the possibility of engaging a third party for the purpose of jamming the broadcasts from Luxembourg. The BBC had its monitoring station at Tatsfield check on whether Radio Luxembourg was causing interference to other stations, and Munich and Trieste were claimed to be subject to interference from Luxembourg's high-powered emissions. (The strength of Radio Luxembourg's signals around Europe was such that the station earned a place in the history of radio technology: following some observations in cross-modulated signals, a theory was evolved that a powerful radio signal passing through the ionosphere could interact in a nonlinear manner with a weaker signal.)

Right up until 1938 (by which time the British government had problems of a different sort), diplomatic moves had been made to remove Radio Luxembourg from the air waves for good. This war, which had been waged for five years, had little to do with the rights or wrongs of introducing commercial broadcasting to Britain, nor with competing technologies, studio technology superior to that of the BBC or the state-of-the-art in German super-power technology. Nor even was it to do with jealousy within the BBC of being outdistanced by another station.

It was to do with imperialism and sovereignty. For the second time in nearly 900 years, Britain's sovereignty had been invaded and violated by another country. The government's monopoly of contact with its citizens was threatened by media from a foreign source. A new weapon had been forged, which through the medium of super-power broadcasting could cross frontiers, bypass border posts, and without censorship reach out to the people in the privacy of their own homes.

But the British government and the BBC were not alone in envying this new temple of technology. The three 600 ft masts on Juglinster Plain had opened the eyes of many to its potential for propaganda broadcasting. Hitler was also casting a covetous eye, and by June 1940 the radio station was captured intact by German forces.

Between June 1940 and September 1944 Radio Luxembourg was part of the Reiche Rundfunk, carrying propaganda to England. In September 1944 the radio station changed hands for the third time in four years, when it was captured by the US 12th army acting under the orders of the US Office of War Information (OWI). As before, great care had been taken to ensure that it was captured intact, and in working order. Within 24 hours Radio Luxembourg was broadcasting propaganda in the opposite direction, towards Germany. Under the control of the US Army Corps of Engineers, it was intended to create confusion and terror in the hearts of the fleeing German civilians, and sought to disrupt the German armies by broadcasting false information.

Table 7.1 *French broadcasting and Radio Luxembourg*

1920	French broadcasting begins with several radio stations
1920	The Amis de TSF (Telegraphic Sans Fils) builds a 100 W transmitter, housed at Rue Beaument, Luxembourg
1920-26	TSF amateur group broadcasting regular programmes to the Grand Duchy of Luxembourg—1000 square miles
1928	Radio Toulouse gets a 8 kW transmitter, old 3 kW sent to Luxembourg
1928	Henri Etienne establishes Societe d'Etudes. Radio Normandy broadcasts programmes of light entertainment, mostly dance music, interspersed with advertisements, after midnight

1928-33	British listeners enjoy these programmes from France, as a welcome break from the BBC's transmissions
1931	Henri Etienne canvases support for commercial broadcasting. BBC is approached but declines to take part
1932	BBC makes diplomatic representations against the proposal
1932	Radio Luxembourg is given a charter from the French government to establish a 100 kW transmitter
1932	The Societe d'Etudes is replaced by a more powerful group with financial backing for commercial broadcasting. This new company, Compagnie Luxembourgeoise de Radiodiffusion, establishes its offices in Luxembourg and buys the plateau of Juglinster plain
1933	French broadcasting of popular music finds favour with listeners in England. The BBC and the British government grow very concerned, with their monopoly being threatened by a handful of French stations from Normandy to Toulouse
1933	International Broadcasting Company (IBC) formed in London for purposes of selling air time on French stations
1933	Radio Luxembourg comes on 1250 metres with 10 kW
1933	BBC monitoring station logs very strong signals on 1185 metres. This log entry turned out to be something of an understatement; Radio Luxembourg was broadcasting with 200 kW carrier power, equal to 800 kW peak power
1933	Radio Luxembourg moves to 1191 metres. Very strong signals reported from all over Europe. If it so wished, Radio Luxembourg could blot out any radio station in Europe, including the BBC's station at Daventry
1933	From May onwards Radio Luxembourg broadcasts 7 days a week: Mondays to Italy, Tuesday Belgium, Wednesday the citizens of Luxembourg, Friday and Saturday for France, with Sundays reserved for British listeners (deliberately coinciding with the BBC Sundays of religious programmes)
1933-38	The BBC, supported by the Postmaster General and the government, impose commercial sanctions and lobby for international support, with the object of forcing Radio Luxembourg off the air
1938	Munich crisis intervenes

Part 2: A weapon of war

Chapter 8

International broadcasting from 1938 to the early 1960s

When the Second World War began in September 1939, it was a purely European affair, involving France, Great Britain, Poland and Germany (united with Austria) and the Soviet Union. By the time the war ended some six years later it had engulfed every continent, with only South America surviving almost unscathed. Many of the nations tried hard to stay out of the war, but had their neutrality violated abruptly and savagely by the armies of one side or the other. Among those countries that were torn apart were the Arabic-speaking countries of North Africa—Morocco, Algeria, Tunisia, Cyrenaca, Sudan and Egypt—and the Middle East.

During this time, the BBC was increasing its propaganda output, from programmes going out in one language in 1938 to 43 different languages by the time the war ended. To many of those countries targeted by the BBC broadcasts in their own tongue, it was a first experience of being tainted by Western culture through Western technology, an invasion that was to have some far-reaching effects. To many it was their first experience of being spoken to from a box: some regarded it as sorcery, but soon came to like it. The Arabic-speaking nations enbraced it, and the leaders of the future made a note of its potential as a political weapon.

When the armies of the Third French Republic and the British Expeditionary Force collapsed under Hitler's blitzkrieg in the spring of 1940, 350 000 British troops fled to England and Marshal Petain was left to save his country from total disintegration and try to restore some dignity in defeat. In June 1940 Petain went to the studios in Paris and addressed the French nation over its broadcasting networks. His moving speech conveyed the power and emotion in the spoken word; it relied on the basic theme of all life, 'harvest from the bountiful earth', but his broadcast was marred by the new French slogan imposed by the Third Reich: 'Work, family and fatherland' had replaced the Republican trinity of 'Liberty, equality and fraternity'.

Petain was eventually replaced by the Fascist leader Laval, but there was another contender for the leadership of France: Charles de Gaulle. Coming from one of the top professional families, he saw his destiny with great clarity—and his timing was perfect, for Britain needed all the help it could get. Following his escape from Bordeaux, and after some discussion with Churchill, he was permitted to make his broadcast over the BBC's transmitters on 18 June: 'I, General de Gaulle, now in London, call upon all French officers and men who are on British soil to get in touch with me.'

The following day, de Gaulle went to the microphone with a much stronger call: 'Faced by the bewilderment of my countrymen, by the disintegration of a government in thrall to an enemy, and by the fact that the institutions of my country are incapable of functioning, I General de Gaulle, a French soldier and military leader, realise I now speak for France.'

This dramatic call to arms signalled the formation of the Free French army, and began the BBC French Section broadcasting to Europe. From 3 September 1939, when Britain declared war on Germany, strict censorship was introduced in Britain. A cloak of secrecy shrouded military preparations in Britain, extending to the media and the BBC. No unscripted broadcasts took place. Censorship, albeit on a voluntary basis to a certain extent, was nevertheless like a fog over England, and it also extended to foreign news-gathering agencies in London.

One foreign news-gathering broadcasting agency that had set up shop in London was CBS News, of the Columbia Broadcasting System, one of America's largest commercial broadcasting networks. CBS News had honed and developed the skills of using newsmen in the various hot-spots around the world, who were practised in the art of getting their stories back to New York via any SW link which happened to be available.

With the idea of covering the war in Europe as if it were a baseball game in St Louis, CBS assigned its men to London and Berlin. In London the CBS correspondent was Edward R. Murrow, and the CBS office was a mere stone's throw from the BBC in Portland Place. One might have thought that a cordial and friendly relationship might have existed between the BBC and the CBS, but it was never so. The BBC itself was nervous of its own fate, since there had been some earlier discussions in cabinet as to whether the BBC should be taken over.

This nervousness on the part of the BBC permeated through the ranks, and every possible obstruction seemed to be placed in the path of Edward R. Murrow. Requests for anything had to be put in writing, particularly if it was to do with visiting a factory or military establishment—which procedure ensured that when permission did eventually come through, the news was no longer topical.

Murrow had placed his assistant William Shriber in Berlin as part of the original plan to report the war from both sides. In complete contrast to London, Dr Goebbels' propaganda machine ensured that all neutral foreign news reporters were accorded full privileges. They were flown to the battle front in Eastern Europe to inspect defences, photographs were permitted and there were no restrictions on reporting. Meantime back in London, Murrow had succeeded in raising many eyebrows in the BBC when he spoke of his plan to take a roving microphone amongst the working-class cockneys of the East End to get some first-hand impressions of the war and the German raids on the London docklands. The idea of consulting the British working man on any subject was not the method of Sir John Reith.

Murrow developed an acute perception of the effectiveness of radio broadcasting, evident in one of his many broadcasts: 'Everyone broadcasts in any language save their own... new and more powerful transmitters are being constructed in order that nation may hurl invective to nation... Radio crosses boundaries and there is no one to inspect its baggage. Propaganda is a legacy of war, and since lying is an attribute of war it is quite natural that the word should play an important part in this war that is going on in the air today. There does not exist, in my opinion, such a thing as a broadcasting station without propaganda on the rights of monarchy or the status quo. We can make propaganda of more tangible things such as cigarettes or automobiles. Individuals may suffer from smoking too many cigarettes or from buying too many automobiles, but these troubles are hardly to be classed with the suffering from the acceptance of an ideal, or a political objective.'

This statement by Edward R. Murrow was the clearest possible reference to the propaganda over the airwaves of Europe of British Imperialism and German Nazism contrasting with the American style of commercial propaganda.

America, from the golden age of the 1930s, had no experience in the art of broadcasting political propaganda: its considerable propaganda expertise was in the domain of the advertising people. This was about to change, however; from 7 December 1941 America was in the war. From being a nation which had only seen the broadcast media as a means of making money, the State Department decided it needed a strong voice on the international airwaves.

The Office of War Information (OWI) came into being, charged with a wide range of responsibilities, including the setting up of a broadcasting arm. The State Department entered into a leasing arrangement with six privately owned radio networks, on a non-profit basis, whereby they would carry a programme feed from a government source. By early 1942 the OWI was broadcasting to Europe, the Middle East, South America and East Asia. The technical and logistical problems of getting the individual stations to operate as an integrated network were immense. The nature and character of the early broadcasts by OWI may not have been as sophisticated as those of the Voice of America ten years later, but they were certainly on a higher moral plain than the psychological radio war that was going on in Europe.

In the Pacific war, the Japanese scored many successes in the first two years, and in this period the Japanese high command showed little interest in the propaganda war, but as the tide of war turned against it Japan entered the field of psychological radio warfare. With technical control vested in NHK and control of programmes vested in the army, Japanese broadcasting went to war. As in the early stages of the shooting war, the Japanese style was an unparalleled success—until 1944, when America copied the style.

Its target was the American GI, and its motive was to make the average American soldier tired of war and killing. This it did by broadcasting high quality, light entertainment recorded by America's top bands, introduced by the sexually charged voices of female announcers, and thousands of GIs fell in love with a voice from which they created in their minds a picture of their loved one. So successful were the Japanese that America realised that, in concentrating on telling the world about American achievements, it had failed in its duty to its own armed forces.

Enlisting the aid of the public relations industry and Hollywood, the OWI set out to better anything that came out of the Tokyo studios. It signed up film stars by the dozen, and was soon broadcasting top-class entertainment, not only to the Pacific islands but also to the Middle East and the European theatre of war. Marlene Dietrich was asked to sing in German, since it was known that German soldiers also listened to OWI broadcasts. Few would claim that she could sing, yet she exuded star quality and could take her audience through emotions from love, joy and happiness to despair; when she sang in German the intensity of emotion was lifted to another dimension. She was the OWI's secret weapon at the end of 1944, when it became clear to the Germans that the war was lost.

Meanwhile in Europe, Allied propaganda broadcasting had assumed a more sinister character, and after D-Day it shifted into a higher gear. Germans, both civilians and troops, were at the receiving end of a barrage of propaganda from the BBC and covert operations. The intent was to create alarm, confusion, despair and terror. As the Allied forces advanced into Germany more and more German-controlled radio stations were turned round as propaganda broadcasters. One top-priority task after D-Day was the recapture of Radio Luxembourg, and the speed with which it was taken and the circumstances of the operation left no doubt of the importance attached to propaganda broadcasting. Within 24 hours Radio Luxembourg became a potent force, creating terror 'like the monster which had escaped from its Frankenstein'.

Propaganda broadcasting reached giddy heights in the Second World War; but this was merely the prelude to the even more powerful role that it was to play in the Cold War between East and West that followed. Viewed in retrospect, the first phase of the Cold War—when one side beamed radio propaganda to another which did not want the broadcasts and so retaliated by jamming, which was

Figure 8.1 *One of the world's first automatically-tuned SSB transmitters. Racal X7929/TA84 5 kW linear amplifier, 1958*

countered by stepping up the propaganda barrage—seems about as safe as playing handball with live grenades. Radio propaganda broadcasting was elevated to *the* weapon of cold war, equating with the nuclear bomb in the ever-threatened hot war.

The 1950s saw two major events that brought a greater emphasis to international broadcasting: the Soviet Union attained parity with nucler weaponry, and the world's first experimental radio satellite was launched.

The threat of nuclear war promoted an awareness of the vulnerability of some forms of communication, and a resurgence in one form of radio communication—high-frequency and SW radio—due to their survivability in a nuclear war. The 1960s marked a renaissance of SW broadcasting: new components, more powerful tubes and new technologies helped to create the first generation of high-power, fast-tuning SW transmitter, endowing it with a 'Raider-of-the-ether'-like quality.

Chapter 9

British censorship and propaganda, 1939–1945

While Chamberlain's Conservative Government pursued a policy of appeasement and co-operation with Hitler, censorship and propaganda broadcasting played an important role. During this time, Chamberlain's Government supported Nazi Germany to the extent of holding discussions about the fate of another country (Czechoslovakia) which was not permitted to be present at the meetings when its fate was sealed. Also not invited was the Soviet Union, a European superpower that shared a border with Czechoslovakia.

Following these meetings, German troops, with the full approval of Britain and France, marched into northern Czechoslovakia. Among the prizes gained by the Nazis were the largest and best-equipped army in Eastern Europe, the giant Skoda arms factory (one of the largest in Europe) and the Tesla company, a giant industrial and scientific organisation that manufactured items from lamp bulbs to the most powerful radio transmitters. It is worth noting that Tesla is still one of Europe's largest companies in the field.

In the period following the Munich crisis, the BBC in its news broadcasts maintained a code of conduct that was not concerned with keeping people informed. As the only broadcasting authority in Britain, answerable to the government, the BBC had a complete monopoly on news broadcasting. Inevitably, it broadcast news that best suited the government, rather than the people. News was intended to act like a bromide instead of a thought stimulator—some described it as chloroforming the people. From September 1939 Britain was on a war footing, and the only source of war news was the Ministry of Information. The MOI's function was to administer the flow of news to the press and the BBC; the assumption was that news would flow like water from a tap, but, of course, this did not happen.

Although the MOI funnelled news, it was not the originator. News was in three categories: home news, foreign news, and war news. The foreign news came from the Foreign Office, which exercised its own censorship. War news came from the three fighting services, each with its own censor. All this information then passed through the MOI, which was itself subject to scrutiny from the Foreign Office. Thus, before any piece of news reached the studios of the BBC it had been subjected to much screening and vetting.

The BBC broadcast programmes on three networks—the Home Service, the

European Service, and the Foreign Service—and it may be assumed that news was tailor-made to suit the listening audience. For example, during the Battle of Britain in the summer of 1940, the BBC in its Home News broadcast figures each day for numbers of German aircraft shot down, compared to losses by the RAF. The figure for enemy losses was over-estimated in some cases by up to 75% in order to boost morale.

When Churchill succeeded Chamberlain as Prime Minister he turned his attention to the MOI and ordered the department to assume full responsibility for propaganda. He left the director general Sir Frank Pick in no doubt that propaganda was to become a key weapon in the future conduct of the war. Pick knew that Churchill distrusted the BBC and its 'pontifical mugwumpery', and he warned his opposite number in the BBC—F.J. Ogilvie—that the government might take over his department.

One of the key elements in Churchill's armoury of propaganda warfare was Reuters Press Agency. Reuters, like most news agencies, owned no radio transmitters but leased time on communication links. It began life as a news agency using pigeons to carry stock exchange news from Aachen to Brussels in April 1849. Julius (later Baron) Reuter, a German linguist, developed his company into the foremost news agency in the world, developing a reputation for truth, accuracy and honesty that was above reproach.

Since 1926 Reuters had bought air-time on Leafield and Rugby, two of the most powerful long-wave transmitters in the world, and it was over this British Post Office communications network that it transmitted news to almost every country in the world. During the First World War, and again in the Second, Reuters Press Agency was eagerly copied by most countries. The stature of its chairman, Sir Roderick Jones, in terms of influence was equalled by few, his importance in the world of propaganda equalled only by Churchill himself, and as befits a man of such importance Jones lived at Hyde Park Gate.

One of the best kept secrets of both wars was the secret subsidy agreement between the British Government and Reuters Press, signed in July 1938. In return for payments, Reuters agreed to transmit news over its network that had been supplied by the Foreign Office. This fabricated news was intended to deceive nations, both neutral and enemy. The first such doctored news came out of Leafield and Rugby on 22 September 1938, and continued throughout the war years and long after. In May 1940 Reuters received its first payment of £64 000 for 'propaganda services'. Nevertheless, Reuters Press succeeded in projecting its image of honesty and neutrality while taking an active part in the propaganda war.

Another player in the secret war of propaganda was the Ministry of Economic Warfare, a dependency of the Foreign Office, which operated from Electra House, London. During the early years of the war the MEW exercised considerable influence. It spent much of its time dreaming up ideas whereby Germany might be defeated by methods other than the obvious—military warfare—which had failed disastrously with the retreat from Dunkirk. Methods considered included economical, political and subversive. It was this department that had the idea of using pamphlets to spread subversive propaganda to discredit Hitler.

There was nothing new in this idea: leaflets had been used by both sides during the First World War, and the use of pictures in cartoon form for propaganda purposes has a long history. During the Napoleonic Wars cartoons depicting Napoleon as a fool were common in Britain, bolstering national morale. Using propaganda in various forms to discredit an enemy leader is something in which the British have much experience, though not always successful.

The first leaflets were dropped in September 1939, and the process continued right through the period of the 'phoney war'. In effect this was the second stage of propaganda. Until 1 September 1939, when the British government was hoping to appease Hitler, the British press was used to project the impression that Hitler was a man of honour. The most prominent players were Lord Beaverbrook's newspapers. Any material likely to give the slightest offence was rigidly excluded. The *Evening Standard* sacked Winston Churchill from his duties as columnist because of his outspoken criticism of Hitler. Lord Kelmsley, owner of the *Sunday Times* and the *Daily Sketch*, had meetings with Hitler in which he is said to have assured him that Churchill, a journalist and backbench MP, carried no weight in Britain. Up to seven days before the outbreak of war, Hitler was being presented to the British public as a statesman who would keep his word. 'Britain will not be involved in a war' was the *Daily Express's* bold headline.

Less than a month later propaganda leaflets were being dropped over Germany, prepared by the MEW and dropped by aircraft from RAF bomber command. The leaflets showed a complete contradiction to the tone of the propaganda put out before September 1939, when Hitler was described by one Fleet Street newspaper as 'Statesman of the year'. In one leaflet Hitler was depicted as a monster; in another, Hitler and Stalin were shown together in an evil-looking picture.

By equating Hitler with Stalin the communist, the leaflet sought clearly to discredit Hitler in the eyes of German citizens. The leaflets attached no blame on the German people, but rather expressed sympathy for them in their plight of being led by a mad dictator. Leaflets of this nature continued for a period of five months, and came to an abrupt halt when Germany invaded France—the end of the 'phoney war'.

In retrospect, it is clear that this phase of leaflet propaganda was a futile attempt to bring about an internal revolution inside Germany with the object of removing the Nazi party from power. Such attempts at subversion can be successful if the people of a country are confused, frightened, impoverished and demoralised; but none of these factors were present in Nazi Germany by the late 1930s. Hitler had brought full employment and had restored German pride.

While leaflets were being dropped on Germany, leaflets carrying a similar message were printed in Polish and Czech languages for dropping on those countries. The intent here was to promote internal uprising against German occupying forces. From 9 September the BBC's European Service broadcast to the peoples of these countries with the same objective in mind, yet during the period of appeasement the BBC did not carry a single broadcast to either of these countries, even at their darkest hour when Nazi forces marched in.

Nothing about British attempts to bring about internal uprising using pamphlets was made known to the British public at that time. It was an offence for RAF servicemen to be caught in possession of such leaflets, which raises the question of why the government concealed from its own people what it was saying to an enemy.

During the year preceding the war, and throughout the war, the BBC was careful not to upset the government in power. Before the forming of the wartime coalition government, broadcasting time was not given to any opposition party, and there were no open-forum-type programmes. it is possible that, had Britain

1. Jahrgang Nr. 5 — Luftpost-Ausgabe

WOLKIGER BEOBACHTER

"Diejenige Regierung ist die beste, die uns lehrt, uns selbst zu regieren."—Goethe.

Das ist der ganze Unterschied.

„Die große stupide Hammelherde."

"Hitler und Stalin sind nicht etwa Schöpfer des Despotismus, sondern umgekehrt, die Bereitschaft der Massen, den Despotismus zu ertragen, machte einen Hitler und einen Stalin möglich."

New York Times.

„Die große stupide Hammelherde." So bezeichnete Hitler in „Mein Kampf" das deutsche Volk.

Sieben Jahre lang ertrug das deutsche Volk die verabscheuungswürdigste Diktatur aller Zeiten. Soll die Welt deshalb glauben, daß Hitler recht hatte?

Die Drei-Groschen-Revolution.

Vor sieben Jahren gelangte in Deutschland eine Klique sogenannter Revolutionäre an die Macht, die es verstand, ihre Gefolgschaft und fast die ganze Nation mit einem Wust von Schlagworten zu betören. Aber alles was diese Klique mit ihrer Drei-Groschen-Revolution erreichte, war — Zerstörung.

Alles, was Deutschland in der Geschichte groß machte, wurde ausgemerzt. Staatsmänner, Gelehrte und Künstler, die im Nachkriegsdeutschland Großes leisteten, wurden ins Exil getrieben.

Was noch an Wertvollem in Deutschland verblieb, wurde zu einer bedeutungslosen Schattenexistenz verdammt. Selbst die eigene Gefolgschaft wurde mit Phrasen abgetan und — unter dem Vorwand, daß es sich um revolutionäre Taten handle — zu Straßen-Krawallen und Saalschlachten mißbraucht.

In Acht und Bann.

Diese Klique sogenannter Revolutionäre war es, die die Welt aus Deutschland verbannte, und Deutschland aus der Welt. Sie ist schuld, wenn Deutschtum heute gleichbedeutend ist mit Barbarei, und wenn Deutschlands Name in den Schmutz gezerrt wurde. Das deutsche Volk ließ es gewähren, — unfähig oder ungewillt, den Zerstörern Einhalt zu gebieten.

Die Welt fragt immer noch: Ist das deutsche Volk wirklich eine „große stupide Hammelherde"?

Das deutsche Volk muß selbst die Antwort geben.

Kurze Nachrichten.

Am 12. Januar warfen englische Flugzeuge über Prag und Wien Flugblätter ab.

Am 11. Januar bombardierten englische Flugzeuge bei Horns Riff drei deutsche Zerstörer.

Der italienische Botschafter in Berlin protestiert gegen das Anhalten des für Finnland bestimmten Kriegsmaterials in Sassnitz.

Schweden verhandelt mit Amerika wegen einer grossen Anleihe.

Figure 9.1 *Wartime aerial propaganda leaflet: two monsters, Hitler and Stalin, 1939 (before Russia became an ally)*

Roosevelt – Deutschland – Österreich.

Die deutsche Presse hat die Rede Roosevelts vor dem amerikanischen Kongreß am 3. Januar nicht veröffentlicht.

Der Präsident sagte: „Es wird jeden Tag offensichtlicher, daß die künftige Welt auch für Amerikaner eine traurige und gefährliche Behausung sein wird, wenn sie durch die Gewalt,— in Händen einiger Weniger —, regiert wird."

„**Wir müssen weiter denken und uns darüber klar werden, was es für unsere eigene Zukunft bedeuten wird, wenn alle kleinen Nationen ihrer Unabhängigkeit beraubt und in Vasallen viel größerer und mächtigerer Militärmächte verwandelt werden.** Wir müssen an unserer Kinder Zukunft denken, wenn sie in einer Welt leben sollen, in der ein großer Teil gezwungen würde, einen von Militärs eingesetzten Gott anzubeten oder überhaupt keinen Gott verehren zu dürfen, in der man keine Tatsachen mehr hören darf, sondern nur noch Gesetze, und aus der die befreiende Wahrheit verbannt ist."

Achtung! Österreichische Arbeiter!

Im Namen der Arbeiter der ganzen Welt hat Citrine, der Führer der englischen Gewerkschaftsbewegung, erklärt:—

„Unter der Führung seiner Nazi-Diktatoren hat Deutschland den Frieden und die Ordnung der Welt zerstört. . . . Auf diesen Diktatoren allein ruht die Verantwortung für den Krieg!"

Die Arbeiter Englands durchschauen den Schwindel des sogenannten „Nazi-Sozialismus". Sie sind entschlossen, den Kampf gegen das Hitler' sche System der Unterdrückung bis zum bitteren Ende zu führen, für Frieden und Freiheit und Menschenrechte.

Gebet im Dritten Reich.

Lieber Gott, mach' mich stumm,
Dass ich nicht nach Dachau kumm' !

Lieber Gott, mach' mich blind,
Dass ich alles herrlich find' !

Lieber Gott, mach' mich taub,
Dass ich an den Schwindel glaub' !

Mach mich blind, stumm, taub zugleich,
Dass ich pass' ins Dritte Reich !

"Seit Adolf Hitler die Macht übernahm, habt Ihr nur ein Privatleben, wenn Ihr schlaft. Darüber hinaus kann es kein Privatleben geben; denn in dem Augenblick, in dem Ihr erwacht, seid Ihr bereits Hitlers Soldaten."
Dr. Robert Ley, 1937.

"Mit jedem Tag scheint das Leben grauer und schwieriger zu werden, als es früher war."
Dr. Goebbels, 27.11.39.

Erlauschtes.

'Gestern ist Bernhard von der Gestapo verhaftet worden.
'So ein anständiger Mensch! Und weshalb?
'Na, deshalb.'

Figure 9.2 *A leaflet depicting life under the Nazis. Dropped over Germany by the RAF during November 1939*

Figure 9.3 *A crude attempt to depict German soldiers as rapists. Dropped in 1940 by the RAF*

possessed privately owned broadcasting companies as in America, with the freedom to broadcast the real facts of the international situation, the Second World War might just have been averted. As it was, the British nation was on the one hand being fed a daily diet from the media that there would be no war, even while, from 1938 onwards the government was preparing for war. Advertising agencies, in co-operation with manufacturers, worked on producing suitable, patriotic pictures for selling products from patent medicines, cosmetic lotions and perfumes to cigarettes. These advertisements depicted a handsome RAF officer, complete with brevet, and a young woman smartly dressed in a service uniform.

The end of the phoney war period saw a change in attitude towards propaganda broadcasting. Churchill was the architect of two weapons that were intended to bring about the surrender of the Third Reich: the bomber and broadcasting. From the outset it is possible that Churchill had grossly overestimated the power and potential of these weapons, yet the bomber fleets went on to incinerate almost a million civilians, and propaganda broadcasting grew to proportions by the end of 1944 that created confusion and terror on a vast scale.

In September 1941, for the first time, Britain acquired a second broadcasting organisation: the Political Warfare Executive, with Bruce Lockhart as its director general. The PWE had a wide list of responsibilities, mostly connected with 'black' broadcasting to the enemy. The BBC and the PWE complemented each other: the BBC kept to projecting the voice of an independent broadcasting service to the people of Britain over its home service network, and to the enemy and the neutral countries over its overseas and European service. The PWE's task was to operate as a covert broadcasting organisation with its output directed to the occupied countries and Germany. The PWE was not universally approved; the BBC, for one, resented it. Green, the Director General, described 'black' broadcasting as both immoral and corrupting. Nevertheless, the setting up of the PWE was probably the only thing that saved the BBC from becoming itself a part of Churchill's ambitions for a radio propaganda war machine.

The PWE did not possess a transmission network of the likes of the BBC; it used radio transmitters of the low-power variety operated by the Special Operations Executive (SOE) and high-power transmitters operated by HM Government Communications Centre (HMGCC), a dependency of the Foreign Office. One of these transmitters was a 600 kW medium-wave installation, codenamed Aspidistra, in an underground bunker in Ashdown Forest, Sussex, whose existence was one of the best-kept secrets of the war. The PWE also prepared programmes for transmission over the BBC's European Service to France and other occupied countries. For this a fictitious character—Colonel Britain—was created from the real person of Sefton Delmar, in the same way as Lord Haw-Haw was created by William Joyce, broadcasting from Hamburg, who became famous with listeners in England. The voice of Colonel Britain was the voice intended to set Europe ablaze.

The expenditure on broadcasting equipment by the British during the war years was staggering. In the normally accepted sense of war, its importance would not rank higher than expenditure on military weapons; but Britain was not in a position to pursue a military war against Germany, having been evicted from France in May 1940. Seen in this context, it is easier to understand why it assumed the importance it did. Just as the air staff were obsessed with the need to develop a four-engined bomber, so Churchill was fixated with the need to develop a radio transmitter that could not only reach Berlin, but would have the necessary power to overcome any radio jamming of its signal. It had been estimated by the engineering department of the BBC that to achieve this, the signal strength needed to be a minimum of 2 mV/m, and to satisfy this requirement would need a transmitted power of 500 kW.

The outcome of this was the construction of the Ottringham radio station (known as OSE-5). Ottringham came on the air in early 1943, with a capability of radiating 800 kW on either long or medium waves. Along with the PWE's 600 kW medium-wave transmitter in Ashdown Forest, OSE-5 was one of the two most powerful radio transmitters in the world.

According to some sources, the PWE transmitter was used by the BBC at certain times to radiate its European service. This fact adds to the complex web of relationships that existed during the war years between the various government departments involved with one or more of the top-secret activities in the world of resistance movements, subversive broadcasting, 'black' broadcasting, phantom broadcasting (assumption of a false identity), information broadcasting and disinformation broadcasting.

As a result of a massive media exercise during the war, the word 'propaganda' took on a new and sinister meaning. The British public were educated to associate Britain with truth and the enemy with propaganda, concealing the origins of the word with faith and truth. Most other countries in the world called their information departments ministries of propaganda, in Britain it was the Ministry of Information. Even today, the BBC World Service is sensitive about the word propaganda in describing its information broadcasting. Many of its staff shy away from the word, believing that this is something the other side does.

9.1 'The biggest Aspidistra in the world'

Before I was posted to Crowborough, shortly after D-Day, I had no inkling of the existence of Aspidistra, even though I had been involved with communications and radio broadcasting; a fact that speaks volumes on the closely guarded secret. It was quite normal in the war years for rumours to abound, but in the case of Aspidistra there were none.

Aspidistra's origins almost certainly go back to the early stages of the war. Some sources suggest that its origins go back to 1938, when Sir Campbell Stuart wanted to set up a propaganda unit to create and distribute false information to the enemy. Something like that was already happening—this was the Reuters secret deal with the Foreign Office to transmit false news over the Post Office transmitters at Leafield and Rugby.

By 1940 the situation had changed; most significantly, the British Expeditionary Force had fled from France, and militarily, Britain was temporarily out of the war for all practical purpose. One source says that it was Richard Gambier-Parry of section VIII of the Secret Intelligence Service (SIS) who proposed building a huge shortwave transmitter, which would resemble a 'raiding dreadnought of the ether'. This may well be true, but it is also a fact that Churchill himself, after the Dunkirk fiasco, came to realise that there were only two possible weapons with which he could continue the war into the heartland of the Third Reich; the

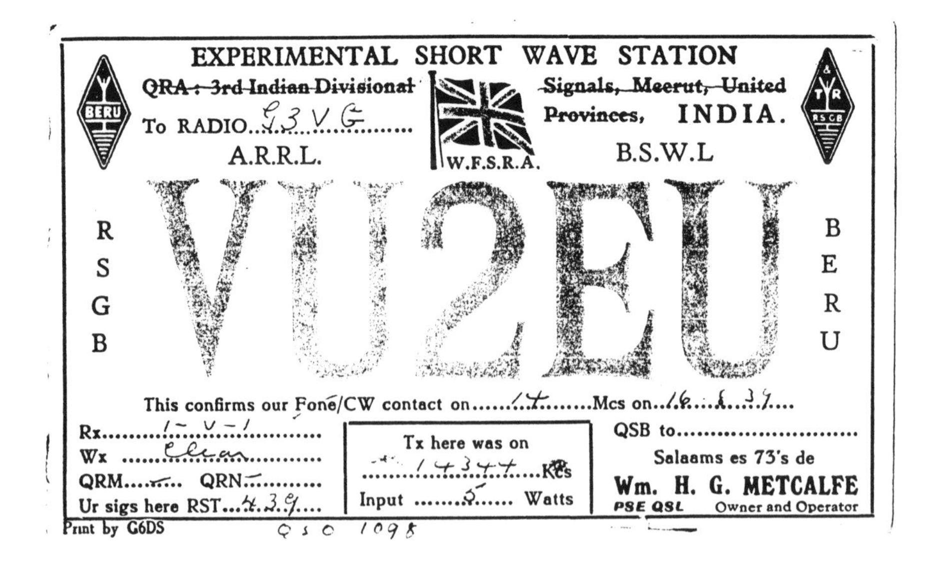

Figure 9.4 *The last days of peace. Short wave communications between G3VG, England, and U3QD, Russia, VU2EU, India, VK3CZ, Australia, and W9ZDS, America*

Radio G 3 VG Was pleased to meet you
on Aug. 8. 1939 at 0530 GMT on the 20 mx. band.
Your signals were QSA 5 R 8 T 9.
Transmitter here uses crystal control.
~~53~~ 6U6G oscillator. ~~46~~ 807 doubler. ~~802~~ 809 buffer. 2 800's in P.P. P.A.
Receiver is ~~O–V–1~~ 8 tube super.

73

A. INGHAM BERRY,
"Pitlochrie,"
15 Kembla St.
Hawthorn, E.2,
Melbourne, Victoria, Australia.

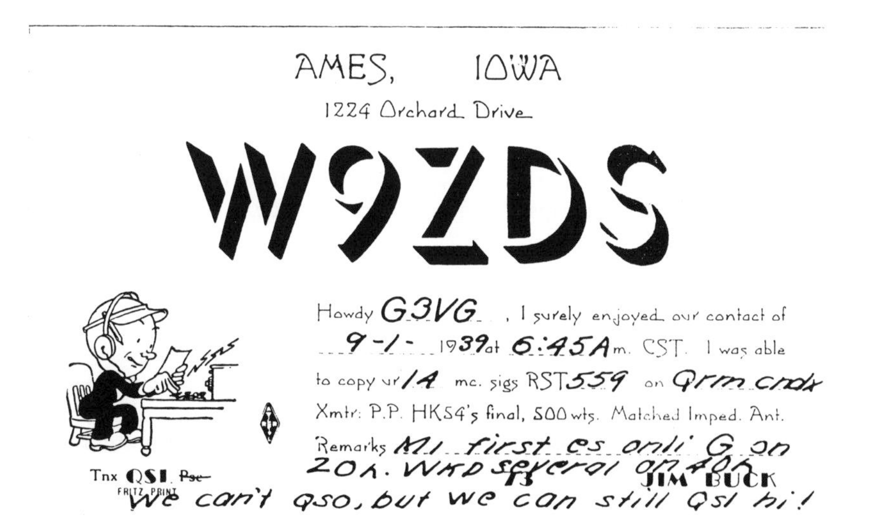

AMES, IOWA
1224 Orchard Drive

W9ZDS

Howdy G3VG, I surely enjoyed our contact of 9-1-1939 at 6:45 Am. CST. I was able to copy ur 14 mc. sigs RST 559 on Qrm cndx
Xmtr: P.P. HK54's final, 500 wts. Matched Imped. Ant.
Remarks Ml first es onli G on 20 m. Wkd several on 40 m.

Tnx QSL ~~Pse~~ 73 JIM BUCK

We can't qso, but we can still Qsl hi!

FRITZ PRINT

bomber, and propaganda broadcasting on a scale to surpass anything that had gone before.

At the beginning of the war Britain was pitifully weak in modern communications equipment, transmitters and receivers: its armed services were equipped with outmoded radio receivers, in many cases fed from lead-acid batteries. Some indication of the under-equipment of the services is that the government bought about 7000 National HRO communications receivers from America, as used by American amateurs. The idea of Britain being able to build a 600 kW transmitter was quite out of the question.

However, American radio companies could. RCA had already built a 500 kW MW transmitter for station WLW in Cincinnati back in 1934. Powel Crosley, who owned Crosley Radio, built and sold cheap radios in the USA; according to reports, these cheap receivers weren't all that good at picking up stations, so Crosley built a 500 kW transmitter, to give them a station they could not miss—so the story goes. The Crosley 500 kW transmitter was certainly the inspiration behind Aspidistra. Incidentally, the origins of the codename Aspidistra are uncertain, but probably taken from a song made famous by Gracie Fields: 'The biggest aspidistra in the world'.

Before the entry of America into the war, at the first top-level meeting between Churchill and President Roosevelt, the leaders agreed to maintain the spirit of revolution in occupied countries, and the organisation of subversive movements, and to undermine the Germans by 'air bombardment, blockade, subversive activities and propaganda'.

With this undertaking signed, the way was clear for Britain to obtain strategic items, including the 600 kW transmitter. This arrived in 1942, its location having already been decided. HMGCC had an underground bunker built deep in the heart of Ashdown Forest; the nearest town was Crowborough, Sussex. If one wanted an ideal site from the secrecy aspect then Ashdown Forest was the place: unless one travelled on unmarked roads, there was little evidence of its existence. Here the Canadian army corps of engineers constructed the underground bunkers behind bomb-proof doors.

WLW in Cincinnati and Aspidistra were similar: the design consisted of three giant output stages which were combined to realise 500 kW of carrier power, which when fully modulated generated an enormous 2000 kW peak envelop power. The three transmitters each generated 170 kW, but in the case of Aspidistra were overpowered to give 200 kW each. The installation became fully operational in 1943. The three plate modulation transformers were almost as tall as a double-decker bus. Occasionally these modulation transformers blew up, and replacements had to be brought in on giant multi-wheeled trucks.

Aspidistra was manned by civilian and service technicians under the control of Robin and Whatton—both Foreign Office employees working for HMGCC. The programme feed came from Wavendon, Bucks. Aspidistra performed a variety of tasks, including propaganda broadcasting, and black broadcasting with intent to cause subversion, alarm and terror among the German forces in France. After D-Day it took on other functions, which included assuming the identity of Radio Calais. (In times of peace this would be against international regulations, since Radio Calais was a properly licensed broadcasting station.) Assuming the identity of another station involves cuddling up to the other station's wavelength, sychronising frequency and winding up the power. The listener's radio receiver does the rest, automatically lowering the sensitivity, so that the real station is drowned out by the imposter. At 0540 hours on 6 June 1944—D-Day—Aspidistra came up on full power to flash the news across Europe, trying to set alight the spirit of resistance.

For all the secrecy attached to its existence, such that it was an offence to even ask about its purpose, the station did suffer from a phenomenon that exists at all super-power stations: audible sounds that came out of the air in the vicinity of the antenna or its transmission line. On a particularly wet day the signal could be rectified simply by stray pieces of transmission line, and the announcements would be audible and clear: 'Voice of Europe'.

The full extent of its contribution to the defeat of the Third Reich is in Foreign Office papers that have not yet been released, but at different times the station was accessed by the BBC, PWE, PWD and ABSIE. After the end of the war the station continued under the control of HMGCC, carrying propaganda from the Foreign Office. In the late 1970s the site was abandoned as being less than ideal from a technical standpoint. The transmitters were moved to Orfordness, a new site on the coast of East Anglia, which provides a better propagation path to Europe by virtue of a better ground wave from the sea.

Although Aspidistra and other stations performed a vital part in creating terror, they were less successful in breaking the will of the Wehrmacht. This is evident from the battle for Calais. The beleaguered and isolated garrison fought on for several months, despite the fact that it was short of food and being subjected to an intense barrage from Aspidistra just across the channel.

On certain transmissions Aspidistra identified itself as 'Soldatensender Calais'.

Chapter 10

German broadcasting under the Nazis

In 1934, when the Nazi party staged a putsch in Vienna, the first move was not towards the government buildings but to the radio station and its studio, from which they broadcast their version of truth to the world. This was the first practical demonstration of the newly acquired strategic status of the radio transmitter. It was a pattern that was to be repeated over and over again in the war in Europe that was to follow.

A similar incident happened some six years later on Hitler's occupation of Denmark. On 7 April 1940, General Kurt Himer arrived in Copenhagen in civilian clothes; his purpose was to reconnoitre the capital and make plans for its seizure. He made arrangements with the port authority for the German cargo steamer *Hansestadt* to dock a few days later. Unknown to the port authorities, that vessel contained a powerful medium-wave transmitter—not domestic electrical appliances as stated in the manifest.

A few days earlier, a German major had arrived with a small force of soldiers, dressed in civilian clothes. They were to await the arrival of the ship, which also contained some small arms. This force—and the radio station—were all that Hitler considered necessary to capture the port and the royal palace, and to secure the occupation of Denmark.

The plan could not have gone more smoothly. After the ship had docked, at midnight, the freight consignment was loaded onto a truck, which was then driven to the Citadel—near the royal palace—and connected to a convenient power point. At 4 a.m. the royal palace was forced by the small detachment of German troops, and disconnected from the outside world. At 6 a.m. the radio station came on air, tuned to the same frequency as Radio Copenhagen, before the real Radio Copenhagen started its daily broadcasting. The bogus Radio Copenhagen made repeated announcements of the King's surrender. Though a small amount of fighting did take place the country was soon back to normal and gave no trouble during the four-year occupation. Denmark was the model for other occupations.

Since the mid 1930s, Germany, with its influence in the world greatly enhanced by Adolf Hitler's rise to power and his acceptance by both America and Great Britain as the statesman of the decade, had become accepted by the civilised world as a leading world power. German technology was seen as some of the world's best. Nowhere was this better illustrated than in its broadcasting equipment. As an overt symbol of the power and prestige that Germany then held, a powerful SW broadcasting station was constructed at Zeesen near Berlin. From 12 state-of-the-art SW transmitters, and a number of high-gain antennas, German propaganda was beamed to all parts of the world: North, Central and South America, Europe, North Africa and the Far East. The programmes were broadcast over all transmitters, which had call-signs ranging from DJA to DJR. They concentrated on two main themes: positively, on the achievements of the new Germany, and negatively, warning the world of the dangers of Bolshevism.

Although Great Britain was not considered by Germany as its natural enemy, the Third Reich in its foreign broadcasting service paid a lot of attention to ensuring that its broadcasts could be heard in England on several frequencies. Conceived in 1938, a new super-power radio transmitter was constructed at Osterloog, Norden. It had no more than 100 kW carrier power, but the antenna system multiplied this by several decibels. This consisted of two mast radiators 150 m high, and eight reflector masts each 120 m high. Its location was so planned to take advantage of a nearly all-sea path to England. (The British Government must have cast covetous eyes on this station for a long time, because it was seized by Allied armies in 1945 before it could be sabotaged. It was later turned round as a propaganda station to broadcast to Berlin during the Berlin blockade. For this purpose the station had its power increased and its broadcasts redirected.)

In 1939 Germany increased its broadcasts to England. Part of this drive was for Reische Rundfunk to recruit a head announcer for its English Section. The post was duly filled by William Joyce who travelled to Germany in September. On the face of it, Joyce was ideal for the appointment. He was born of upper-class parentage, public-school educated, extremely articulate and fluent in German.

From late September Joyce broadcast to England from studios in Hamburg. In his distinctive upper-class accent, Joyce went to work reading out English news on the conduct of the war. During the period of the phoney war this consisted of listing shipping losses, while the British Expeditionary Force was sitting around waiting for something to happen. It did, when the German panzer divisions swept through the lowlands of Western Europe with a speed that left the British army breathless and bewildered. A new word, 'blitzkrieg' found its way into the English language.

Much has been written about Britain's finest hour: myths have become facts and legends spun on the evacuation of troops from Dunkirk. There has been less mention of the vital installations that the Wehrmacht captured intact and in full working order, including power stations, factories, shortwave communication links, and high-power broadcasting stations and their studios.

By the end of the summer of 1940, Germany had seized control of almost every broadcasting station in Europe, a chain of radio stations that extended from Tromso in the Arctic Circle through Holland, France, down to the Mediterranean coast and to the shores of North Africa, while in Eastern Europe the powerful senders of Radio Prague, Radio Warsaw and Radio Belgrade were already under the yoke of the Third Reich. At this time Germany possessed the largest broadcasting network in the world. This may have been a contributing factor to Churchill's growing obsession with the power of radio broadcasting as a propaganda weapon to be turned against the Nazis, in the hope that it could destabilise the Third Reich and bring about its collapse.

Yet Hitler seemed to see radio broadcasting in a different light. For Hitler the power of radio broadcasting was in its role as an uplifter of morale. The object of occupying countries is for economic gains. These are best achieved by the restoration of law and order, stability and the bringing about of good relations between the occupied and occupier. To this end, each occupied country remained in day-to-day control of the national radio stations with the minimum of interference from the military governors—with one notable exception, the Grand Duchy of Luxembourg.

A 5 kW station would have been quite adequate to cover such a tiny State, yet it possessed the most powerful radio station in Europe. From its early origins Radio Luxembourg had been conceived to serve the countries of Europe with propaganda of the commercial variety; the selling of patent medicines, goods and holidays abroad. Now it was to be used to sell Nazism to the British.

Under pressure from the British Government, Radio Luxembourg had closed down on the outbreak of war. Purportedly this was because it could be used as a navigational aid to German aircraft, but a more likely explanation is that the BBC had wanted to silence it since 1934.

The Germans had taken control of the rest of the IBC network, Fecamp and Radio Normandy, but Radio Luxembourg, with its massive radio masts on Juglinster Plain, was the jewel in the crown. From June 1940 the station was under new management by the corps of engineers of the Wehrmacht. It still retained its own studios in Villa Louvigny, but now it had a parallel programme feed from the Hamburg studios of Reische Rundfunke. Along with the radio station and its studio the Germans secured intact the valuable archives of high quality gramophone records, containing the best of American light entertainment.

By the summer of 1940, William Joyce's voice could be heard all over England with his typical announcement: 'Germany calling, here are the stations: Calais One, 514 metres, Calais Two 301.6 metres, Koln 456 metres, Breslau 316 metres, Luxembourg 1293 metres and on shortwave 41.27 metres'. This was followed by war news in English, making it painfully clear that the war was not running in Britain's favour.

The British Government viewed the situation with alarm; it knew that much of what was said was true, and Britain was losing the war. Joyce was a serious speaker, not given to excitement or rhetoric and British people did tune in to his broadcasts from Radio Luxembourg. After some months of anxiety, the problem seemed to solve itself when Jonah Barrington, a popular columnist of the *Daily Express*, christened Joyce as 'Lord Haw-Haw'. The tag was an instant success. The article painted Joyce as a clown of the upper English class, inviting the working class to ridicule its social superior—with the full blessing of the government and a feeling of patriotism.

From the serious, well educated person that he was, Joyce had his status reduced to that of a music-hall turn. Within a few days after the *Daily Express* article, Lord Haw-Haw became a topic for discussion in working men's pubs. He was even imitated by comics on the stage and on the BBC. People tuned in to Radio Luxembourg just to hear his voice; his credibility had been destroyed.

The destruction of the status of any person who represents a threat to the state is an art in which the British are skilled. Although the Lord Haw-Haw tag was attributed to the *Daily Express*, it is almost certain—given the threat that he posed—

that the name was assigned by a government department and leaked to the press in the time-honoured way.

Nevertheless, Joyce continued to broadcast to listeners in England without once changing his impeccable style of delivery, and privately the government was still worried. Possibly as an attempt to discredit Joyce even further the press ran stories such as 'The more people who tune in to foreign propaganda broadcasts, the greater the joy and laughter'. This was a replica of the technique used to good effect against Hitler, given the title Corporal Hitler, and a century earlier against Napoleon. In the invasion plans for Europe in 1944, two tasks were given a top priority: the seizure of Radio Luxembourg with minimum possible damage, and the capture of William Joyce, unharmed, so that he might be tried for high treason.

10.1 Lilli Marlene

The 'Lilli Marlene' broadcasting station of the German Afrika Korps became the most famous propaganda radio station in the Middle East. Its nightly broadcasts forged a bond between the two opposing armies. The cement of this bond was the type of music and the opening number played by the station. Night after night the soldiers of Germany and Britain listened to the soulful music, made famous by Marlene Dietrich in the film 'The Blue Angel'. It forged a bond between two armies, and for many a British officer it was a nostalgic recall of the decadent Berlin nightclubs of the early 1930s. Lilli Marlene became a legend with the British 8th Army, in much the same way as American marines in the Pacific war fell in love with the ethereal 'Tokyo Rose'.

Yet the origins of this radio station, and others such, had nothing to do with propaganda broadcasting as an instrument of war against an enemy. It began in 1937, the period of growth of the German war machine. Conscious of the use of broadcasting for the maintenance of morale, the German high command placed a contract with the Telefunken company to develop and produce a small number of high-power mobile broadcasting units to serve the Wehrmacht. One design was for a 20 kW MW transmitter, another for a 50 kW LW transmitter. Given the state of the art at that time it was quite an achievement to accommodate such high power transmitters in trucks. Each broadcasting station comprised a number of four-wheeled trucks, each containing a stage of the radio station: drive unit, radio-frequency amplifier, output stage, modulator stage and mobile power generator.

The radio stations were officially designated 'Lebenslauf des Sender Fritz' and 'Lebenslauf des Sender Heinrich', and served with the German army as entertainment stations in many theatres of war. It was, however, in the Western desert that these stations made their biggest impact and became legendary. No less remarkable is the fact that although Germany lost the war, these mobile radio stations eventually found their way back to the war-torn, beleaguered city of Berlin and were used as a means of broadcasting emergency news to the Berliners when all other radio stations had been damaged or destroyed.

10.2 Recapture of Radio Luxembourg

Thirteen weeks after D-Day, Radio Luxembourg was liberated from its German conquerors under exceptional circumstances. Despite the efforts of the departing Wehrmacht troops to destroy it, the studios and the transmitter were captured intact by a special task force of the American 12th Army Group, acting under the direct orders of the Psychological Warfare Division (PWD). For the second time in the space of the war years the super-power station changed hands without suffering damage. Within a few days after its recapture it was on the air again, this time as an Allied propaganda station under the control of PWD.

As the advancing Allied armies penetrated into Germany, radio broadcasting stations were captured one by one and were turned round as quickly as possible into propaganda stations to harass the routed German forces. In the east a similar thing was taking place following the breakthrough of the Russian divisions, so that by the last month of the war only two German radio stations remained in operation: Berlin and Hamburg. Hamburg resisted all efforts to capture it, and the radio station continued to broadcast news—often tragic—right up to the end.

On 4 May 1945 the historic announcement was heard from Hamburg: 'Hitler is dead. . . the iron curtain from the East is moving nearer'. On 8 May it announced the surrender of Germany, one day earlier than that broadcast by the Allied Command. Joyce, who towards the end had made impassioned pleas for an end to the war, made one last attempt for an end to the fighting in the west, warning of the dangers of communism from the east. His words were to be echoed in the next two years. Joyce was captured in the studio of Hamburg Radio and taken into custody to await trial.

10.3 Philosophy of Nazi broadcasting

Nazi propaganda combined traits common to other nations' war efforts with a demonology and an idealism that made sense only in the context of German history, and particularly with the humiliation it had suffered as a result of its capitulation in the First World War.

During the Second World War, Nazi propagandists made a massive effort to explain and justify the war effort to its people. Nazi ideology was adapted to suit needs, and employed mass-media techniques to reach the entire German population. It made vast use of books, brochures, window displays, placards and a magazine *Der Propagandist*, published by Gau Propaganda Central Office (GPL). Leaflets were used on a massive scale, and, of course, there was the German State Radio Broadcasting Authority, which came under the administrative control of the Reich Radio Society.

The Reich Radio Society managed the German radio network. Its programme division co-ordinated all broadcasts and arranged for exchanges between the 26 radio stations composing the Greater German Radio. There were 13 regional stations and numerous local outlets, which latter did not broadcast on their own, but carried programmes for the regional stations, and for the national transmitter *Deutschlandsender*.

Hans Fritzsche was appointed head of the Reich Radio Society in 1942. Under his direction, an overseas-broadcast monitoring service (*Erkundungsdienst*) was set up to monitor enemy propaganda transmissions. German broadcasting could then act quickly to counter the enemy propaganda.

Subversive propaganda broadcasting from Britain was a menace that often reached crisis proportions, when these broadcasts from England pretended to emanate from within Germany, with broadcasters such as 'SA man Max Schronder' broadcasting anti-war views. Before the war, the German industry had produced a 'people's radio receiver', *Volksempfanger*; there were rumours in Britain that this radio set had been designed to be incapable of receiving enemy broadcasts, but there was no truth in this story. Much as Hitler would have liked this to be so, technically it would have been difficult. Although it was said to be a crime to listen to enemy broadcasts in Germany, there seems to be no evidence of any prosecutions, and it would seem that propaganda minister Joseph Goebbels placed much reliance on German patriotism to act as a deterrent. As the war in Europe developed, and the German army suffered reversals, the numbers of Germans listening to enemy broadcasts began to increase.

Nevertheless, there was never much danger of the enemy broadcasts corrupting the minds of the German nation, because of the powerful style of propaganda created by Dr Goebbels. Among the Nazi leaders, Goebbels was the intellectual; it has even been said that he was the only true Nazi, surrounded by more than a few gangsters. His success as a propagandist lay in his passionate belief in his task. Goebbels was well educated and well informed; he knew the origins and meaning of the word 'propaganda', and to him it meant propagating the sacred truth as in the Christian 'sacra congretario de propaganda fide'. So great was his passion to uphold the German nation that he forbade the use of the word 'propaganda' by the media in describing the lies told in enemy broadcasts; the word must only be used in the context of German broadcasting.

The Goebbels propaganda machine was efficient and it had only one target: the entire civilian population of the Third Reich, not the people of the occupied countries, nor the Wehrmacht. His strategy could not be faulted: the German army could take care of itself and its morale was always high, but the future of the entire German race hung on convincing the nation that it was the greatest on earth.

To combat the attempts by the enemy to destroy the morale of the German people, Goebbels was faced with a difficult task; he had to add to the character of the German nation, thereby promoting even greater courage from the people in the face of growing hardships, adversity, and danger from Allied bombing of German cities. So successful was he that even when the major part of the German forces had been routed, killed or captured and defeat was inescapable, the remaining German forces fought even harder, and the output from the German munition and aircraft factories was at its highest.

But even Goebbels, the supreme master of the art of propaganda, is said to have admired the talents and skill of Churchill, who turned the Dunkirk defeat into a heroic myth, and thus recreated the will to fight on when all seemed lost. Goebbels used this technique to prepare Germany for its own Dunkirk—the defeat of its finest armies at Stalingrad. As the war dragged on through 1943 and into 1944, with the tide of success going against Germany, the tone of propaganda broadcasting to the German nation did not change.

As well as lauding German success and German virtues, of course, it was necessary to build a common sense of hatred of the British. The anti-British broadcasts were based on the theme and the history of the British empire, dwelling on the concentration camps of the Boer war and atrocities committed against the

Irish. Although historically accurate, these broadcasts lacked rhetoric, intensity and passion, and because of this German foreign propaganda broadcasting was not always as effective as it might have been. Enthusiasm, intensity and passion were reserved for national propaganda to the German peoples, it was positive, real and came from Goebbels' heart. Because his belief was a real one, his style of propaganda could not simply be changed or turned off like a tap; it was a continuous process, automatic and repetitive.

As the war progressed to its conclusion, the intensity of propaganda increased. Goebbels and Hitler were both of the same mind: that it was the German people who mattered, and any other application for propaganda broadcasting, such as the type used by the Allies, was secondary to the need to bolster the morale of the nation. The roots of this philosophy, like much of Nazi philosophy, lay in 1918. The collapse of Germany's capacity to fight on occurred quite suddenly, taking its enemies by surprise. The cause was directly attributed to the rapidly accelerating breakdown of the German people's will, rather than the German army itself. Hitler served in the First World War, and the shame of the demoralised high command surrendering when it could have fought on was the spur that drove Hitler and Goebbels on to make sure that such a shameful thing did not happen a second time—backed, of course, by the Allies' insistence on nothing but an unconditional surrender.

Therein lies the tragedy of the Second World War. So successful was the Nazi propaganda executed by Goebbels and Hitler that even when defeat stared Germany in the face, it refused to believe it.

Despite the differences between subversive propaganda aimed at an enemy and the uplifting type aimed at one's own people, there is a similarity in that both aim at the conquest of the mind. For propaganda to be effective the message has to be repeated until eventually it is perceived to be truth. The real danger comes to those who, in propagating their faith, come to believe it to be the absolute truth. That is what happened to Hitler and Goebbels. In the passionate belief that right was on the side of the German nation, it was a short step towards believing that right would, in the end, prevail with the aid of God.

Chapter 11

US wartime broadcasting

In early 1941 the US government instructed the Federal Communications Commission to set up a special radio division for the purpose of listening to foreign shortwave broadcasting stations, and analyse this intelligence information. This department operated only in a small way at first, until the Japanese attack on Pearl Harbor and America's entry into the war, which had now acquired a global perspective.

The Foreign Broadcasting Intelligence Service (FBIS) set up four SW monitoring receiver stations: at Portland, Oregon, Kingsville, Texas, Guildford, Maryland, and Santurce, in Puerto Rico. Using sensitive SW communication receivers in conjunction with directive antennas, selected foreign broadcasts were recorded on discs which were kept for archival use. Selected material was then transcribed and translated. The different US war departments received weekly reports, plus special reports such as 'Radio Tokyo's racial propaganda', 'Underground movements and morale in Japan', 'Berlin's claim to Allied shipping losses', 'Nazi portrait of the American soldier' and 'Reactions to the bombing of Japan'.

The broadcast transmissions monitored by the FBIS included foreign neutral broadcasts from many different countries, but more particularly the official enemy-controlled broadcasts, and the unofficial radio stations broadcasting propaganda, counter-propaganda and covert broadcasts—i.e. radio stations that were not what they claimed to be.

The FBIS also tracked many SW broadcasts emanating from places within the USA; and these were of special significance for reasons of internal security, since the US had right-wing fascist movements within its own borders. A station calling itself 'The radio station of all true Americans' was heard in the USA from early 1942; its broadcasts were received most evenings from 8.30 onwards. This station opened its transmissions with a rendition of 'The stars and stripes', and signed off after its news programmes with the National Anthem. It attacked the character of President Roosevelt, and carried news items intended to stir racial tension, and to shock the nation. One broadcast reported that a military air transport had landed at Hoboken field with a load of American army nurses made pregnant by American officers. This news report was aimed at the families of nurses and at wives of American officers overseas. It also found a target among certain officers and enlisted men, whose perception of the duties of an army nurse varied from that held by nurses, most of whom joined the army out of patriotism.

The 'Radio station of all true Americans' contrived to build up credibility as a genuine US SW station by such ploys as referring to difficulties with the FCC and its transmitting licence over its 'frank disclosures' of the behaviour of officers, such as the story mentioned above, and others that were equally forthright about US officers' brothels. On 10 June 1942 it triumphantly announced that the FCC had agreed to restore its licence, but on another wavelength.

This propaganda radio station posed many problems to the US authorities, not so much in the size of its listening audience but more because of the way the rumours multiplied and gained some credence in the process. In common with all the best propaganda stations, it relied on a grain of truth in the stories and news reports it put out, and it was probably this that gave reason for concern. This radio station contrived to broadcast regularly, and many listeners gained the impression from its broadcasts that it was located in Maine. Eventually, the FBIS reportedly traced the broadcasts as emanating from Germany with the aid of long-range HF direction-finding equipment. The station remained operational until late 1944, when its broadcasts ceased as abruptly as they had started.

Given the regularity of the broadcasts, and the strength and consistency with which the transmissions were received, plus the fact that the Nazis frowned on covert broadcasting, a more likely explanation may be that the station was in fact a covert anti-American radio station located in America, and quite probably in the mid-Western region of the USA where large numbers of Germans had settled in the early 1900s.

FBIS did successfully track down some propaganda radio stations that were actually located within the USA: one, a SW station that regularly signed off its transmissions with 'Heil Hitler', was found in Peoria, Illinois. Some of these propaganda broadcasts exploited the fact that the US armed forces exercised a policy of segregating white and non-white enlisted men, in line with government policy towards coloured persons. Towards the end of the war, attempts were made to tone down such attitudes.

All leading participants in the Second World War became involved in 'black' broadcasting to a lesser or greater extent (although the leading exponent by far was Great Britain). The FBIS kept track of over 60 such stations, including German-language stations purporting to be a freedom movement within Germany. Such was the cloak of secrecy attached to black broadcasting that even the FBIS did not know the identity of black radio stations operated by the Allies.

As the war developed, the FBIS grew to become one of the largest intelligence-gathering agencies in America; and by the end of the war it numbered several thousand personnel, including those on detachment from the signal corps. Its opposite number in Britain was the BBC monitoring station at Caversham Park, Reading, with which it exchanged information.

Scholars, academics and experts in social studies who worked at FBIS as analysts acquired a nose for detecting to what extent foreign propaganda broadcasts relied on an element of truth in their news items intended to shock, or whether they were a complete fabrication of lies. Yet the analysts must have known that war debases and corrupts every aspect of human behaviour, and that a prolonged and total war brings about this degradation in a slow and insidious way that makes ordinary people unaware of the change. Lapses in moral behaviour are never

one-sided; the Japanese beheaded US soldiers, and the Americans developed the flame-thrower to burn the enemy in his bunker; RAF airmen dropped phosphor and magnesium to burn civilians in German cities, while the Germans went about their business of ridding the world of the Jews. The war effort on both sides depended for its sustenance on atrocity stories calculated to whip up hatred for the enemy. This was something that was fashionable in Britain from the start of the war. Sick jokes about the conduct of the enemy were considered appropriate material to broadcast to Germany. A typical example was the following conversation between two Germans:
1st German: 'I've got a good job, with lots of overtime.'
2nd German: 'What do you do?'
1st German: 'I work for the SS as a gravedigger.'

11.1 Office of War Information

From its earliest days in 1920, US broadcasting was characterised by freedom of speech and was free of government control. The Federal Communications Commission was concerned with laying down technical standards rather than programme standards, and there was no government-controlled broadcasting agency.

After 7 December 1941, the day of the attack on Pearl Harbor, America's attitude to everything was changed. Although successive governments had been slow to realise the value of sound broadcasting as a propaganda weapon, the US Government under Roosevelt took immediate steps to correct previous deficiencies in this direction. However, the problems in the way were immense.

The government as such had neither a news-gathering capability nor a transmission network. These were in the hands of companies such as NBC and CBS. Both broadcasting agencies had developed efficient news-gathering capabilities, with correspondents and reporters covering the war in Europe, skilled in the art of arranging SW relays at short notice from various cities in Europe, providing up to the minute news reports, yet avoiding apportioning blame to either side.

Broadcasting was big business, controlled by business people, and they wanted to stay in business. To this end, both NBC and CBS meticulously employed people who analysed the situation without promoting opinions. The outcome of the war was uncertain, but the odds were that Germany would win. When America entered the war its main enemy was Japan, and this was a war that America seemed to be losing.

Shortly after Pearl Harbor, the Government took steps to take control of external broadcasting, which until now had been in the hands of private companies. The Office of War Information was set up to be a government agency acting under the direct control of the President's office. The OWI was given immediate powers to negotiate operating contracts with those companies that owned SW transmitters, and to set up its own studios to originate propaganda programmes for transmission over the leased SW transmitter networks. The head of the OWI was Elmer Davis; he later appointed playwright Robert Sherwood to head OWI international broadcasting, with John Houseman in charge of programming.

In December 1941 there were six broadcasting companies in the USA that owned and operated the radio stations on a network basis. These were: Columbia

Broadcasting System (CBS), Crosley Radio Corporation (CRC), General Electric (GE), National Broadcasting Company (NBC), Westinghouse Electric (WE) and Worldwide Broadcasting Corporation (WBC). The OWI arranged for the FCC to licence all six broadcasting networks to carry SW transmissions for the Government; at the same time, the OWI made budget requests to Congress for 22 additional SW stations. This was to become the 'Voice of America'.

In the beginning, the technical problems were immense. Much of this was due to the way the private companies had developed their broadcasting networks: each company had tended to a policy of designing and constructing its own short wave transmitters, in some cases built by the station staff themselves. As a consequence there was little or no similarity in parts between the six broadcasting agencies. The only common features between the various transmitters were in the basic design: all were of the 50 kW carrier power, using the classic class C radio-frequency amplifier stages and high-level class B modulator design. All were therefore truly representative of the state of the art at that time, and reasonably efficient.

On Long Island, CBS operated two transmitters in Brentwood before the war: W2XE and W2XDV. To meet the needs of the OWI, CBS installed two more 50 kW transmitters at another site owned by Mackey Radio, also on Long Island. These transmitters, designed and made by Federal Telegraph, an associate of ITT, were unusual in that they had two modulator sections and three RF sections. By this arrangement, two transmitters could be in service at any one time, while the third RF section was prepared for the next frequency change. At the chosen time the change to another frequency was almost instantaneous. Frequent channel change is a characteristic of SW broadcasting, and this method was the only means to make it happen fast: changing coils and implementing a new line up on a SW transmitter would otherwise take up to an hour. To go with these short wave transmitters, CBS had 15 curtain arrays of the directive type, with a manual arrangement for selection and switching of open-wire feeders.

CRC operated a single 50 kW SW transmitter, on the same site as a 50 kW MW transmitter at Mason, Ohio. The antenna was a single wideband rhombic. CRC's studios were in Cincinnati. GE's transmitters were at Schenectady, New York, where they had been since the early 1920s, when it operated stations W2XAF and W2XAD. GE ran a third 50 kW transmitter from this site for the OWI, and another at Belmont, California. GE engineers favoured the use of SW curtain arrays. Its zone of coverage took in South America and the Pacific.

NBC had its studios at 30 Rockefeller Plaza, where it had always been known as '30 Rock'. Its transmitting station was at Boundbrook, New York where it had always been, the home of W3XAL. Here the NBC engineers installed an arrangement of three RF sections and two modulators—like that at the CBS station on Long Island, except that these were designed by RCA. Curtain arrays were used to cover Europe and parts of South America.

WE owned and operated two transmitting sites, one at Boston, which housed a 50 kW transmitter WBZ, the other at Pittsburgh, home of the famous W8XK, where WE engineers put together a second transmitter to carry OWI programmes. Rhombic antennas were used, and were beamed towards Europe and South America.

WBC was owned by the Worldwide Broadcasting Foundation. Its slogan was 'Dedication to enlightenment'. WBC was another pioneer broadcaster with its roots

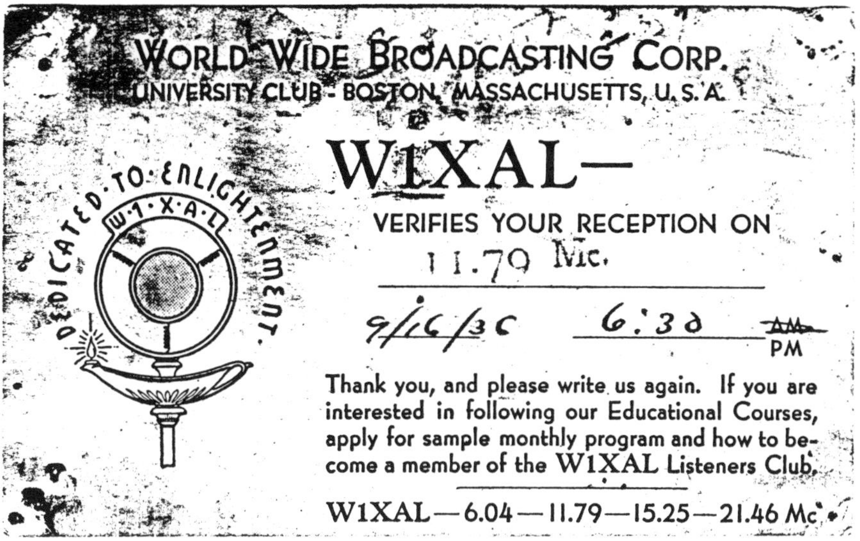

WORLD WIDE BROADCASTING CORP.
UNIVERSITY CLUB - BOSTON, MASSACHUSETTS, U. S. A.
DEDICATED TO ENLIGHTENMENT
W·1·X·A·L
W1XAL—
VERIFIES YOUR RECEPTION ON
11.79 Mc.
9/16/36 6:30 PM
Thank you, and please write us again. If you are interested in following our Educational Courses, apply for sample monthly program and how to become a member of the W1XAL Listeners Club.
W1XAL—6.04—11.79—15.25—21.46 Mc.

Figure 11.1 *QSL card from WIXAL, 1936*

in the 1920s. In the early days, WBC had operated W1XAL. In 1941, for the OWI, it operated two 50 kW SW transmitters, WRUL-1 and WRUL-2. As with the other broadcasting companies, these two transmitters had been built by the station engineers at Scituate, Massachusetts.

With this motley assortment of 11 different transmitters operated by six different broadcasting companies, OWI went to war. Such was the birth of what would eventually become the largest broadcasting agency in the world, the Voice of America.

The master control point for programme collection and distribution was located at the studios of OWI, on the seventh floor of General Motors Building, 57th Street, Manhattan. Programme feeds to the individual transmitter sites scattered between the East and West Coasts of the USA were sent over lines leased from AT&T. For the first time in its history, the US Government was to exercise complete control over a news medium. Although this move on the part of the Federal Government was essential to the war effort, there was much concern expressed on the part of many working in radio that it was a move down the road of propaganda broadcasting, which would ultimately lead to government control over national broadcasting, as practised in Britain, Germany and Russia.

However, the Government decided against nationalising these six companies. The licences would remain with the licensees, but the government would supply programmes. Contracts for installation, operation and maintenance were to be renewed on an annual basis with the six broadcasting companies, who in turn agreed to run the networks on a cost-only basis, so as to avoid any impression of making profit from war. In fact it was a lucrative arrangement that suited everyone.

From a geographical viewpoint, this hastily-cobbled-together network was less than satisfactory. Practically all of the transmitters were located on the East Coast, around New York, Pittsburgh and Boston, where broadcasting had begun. However, the prime target zone for propaganda broadcasting was Japan and the Pacific. By contrast, the West Coast started to expand when San Francisco became the seaport vital to the war against Japan.

The planned expansion of the OWI's overseas broadcasting arm was to move transmitter sites to the West Coast. But 1942 was not a good year to think about expanding overseas broadcasting. The national priorities were for production of guns, tanks and aircraft. Radio transmitters were regarded as civilian items of hardware.

Although its role was to carry propaganda programmes to the world at large, and the enemy in particular, the OWI became enmeshed in another problem: the need to broadcast entertainment to the American forces overseas. The OWI's SW broadcasting network stepped into the breach to carry programmes of radio entertainment for the GIs overseas. But this was not the task for which OWI had been formed, and it required a capability which it did not possess. Eventually another broadcasting organisation was created with this responsibility: Armed Forces Radio Service (AFRS), leaving the OWI to concentrate on its specific role.

When the international network of 12 leased SW transmitters was running, the board of OWI set about its next task, which was to expand the network. In terms of the world league table of international propaganda type broadcasting Britain was at the top and the USA at the bottom; a performance gap that was likely to widen when the BBC implemented its expansion scheme. Already in the advanced planning stages, the BBC intended to bring into service new transmitter sites in 1943, which would make the BBC the largest international broadcasting agency in the world.

Possibly inspired by this rivalry, the OWI implemented more SW outlets. A spare 50 kW transmitter was obtained from station KFAB in Lincoln, Nebraska, and re-installed at Mason, Ohio. Two more dual-channel 50 kW transmitters were obtained from RCA, which enabled four transmitters to be placed into service if the frequency-changing facility was sacrificed. These RCA transmitters were installed in a transmitter station owned by CBS in Long Island. The next step was to set up a brand new station for long-term operations.

As part of the plan to bring the short wave transmitters nearer to the enemy, a new transmitting station was constructed at Hunter's Point, San Francisco. Into this was placed a new 100 kW SW transmitter of GE manufacture, followed by a 50 kW transmitter, also of GE manufacture. The licensee for this new station was yet another company: Associated Broadcasters Inc. Both these transmitters used curtain arrays beamed on the Pacific area, intended to increase signal audibility.

Two identical turnkey projects were then planned to come into service by 1943: new stations at Dixon and Delano, California. Federal Telegraph was awarded a contract to design and build two 200 kW SW transmitters—a very ambitious step in SW technology. The transmitters used new tubes in the final output stage—the design for which had been stolen from France under the eyes of the Germans in June 1940, and a prototype tube brought to the USA.

Dixon and Delano were also equipped with dual-channel 50 kW SW transmitters, and by an ingenious arrangement of switching modulators to the transmitters in use, both transmitter stations could operate on three different frequencies, with 200 kW on one and 50 kW on each of the others. In contrast to most of the other SW stations, Dixon and Delano used rhombic antennas, but these were later augmented with the more popular type of curtain array. From these two transmitting stations—the highest powered in its entire network—the OWI was now able to beam its broadcasts to Australia, to the Spice Islands of the Dutch East Indies (now occupied by the Japanese) and even to Japan itself.

The turning point of the Pacific war was the Battle of Midway. With victory now only a matter of time, OWI was allowed funds by Congress for further transmitters. The Crosley Radio Corporation was awarded a contract to design and construct a new SW transmitter station at Bethaney, Ohio. Larger even than the stations at Dixon and Delano, Bethaney was equipped with three 300 kW class B modulator stages and six 200 kW RF channels, all modulated with the same programme feed. Thus the total transmitted output power of Bethaney could be up to 1200 kW: over twice the output power of the entire OWI network in its original form, which had eleven 50 kW transmitters.

By the end of 1944, the OWI's network was reaching maturity; not only in terms of output power, but in other technical aspects such as frequency assignment, choice of optimum frequencies, transmitter switching, antenna selection, and the programme feeds to the various transmitter stations now dispersed around the USA. As well as the transmitting stations mentioned, the OWI operated stations outside the USA. These included a 50 kW SW transmitter, and two MW transmitters (of 100 kW and 50 kW rating) in North Africa. At different times these carried black broadcasting for the psychological warfare department.

By the end of 1945, OWI operated a network of 30 SW transmitters; but even so, it could not compare with the BBC, which by now had grown into the largest propaganda broadcasting network in the world; 37 transmitters, pumping out 850 hours per week in 45 languages. This output was greater than that of the USA and the USSR combined.

The OWI, and the private broadcasting companies that were the substance from which it was created, deserve considerable credit. Its wartime achievements were instrumental in alerting the US Government that, while national broadcasting can best be performed by private capital, there was a case for building up an international broadcasting agency funded, maintained and controlled by the government to carry the voice of America to the world.

Shortly after the end of the war with Japan in September 1945, along with other war departments the OWI was formally dissolved. Its SW broadcasting network became a part of the state department and preparations were made for a rapid post-war expansion. By the early 1950s America had declared itself the world's super power. The Voice of America was broadcasting ceaselessly to the world. The American presence was felt on every continent both in military might and the broadcast word. Terms such as 'fascist' and 'Nazi' were no longer acceptable terms of abuse now that the Germans were potential allies. Their place had been taken by the equally emotive words 'Reds' and 'Commies'. McCarthyism was born.

In 1953 the United States Information Agency (USIA) came into existence, ranking in importance alongside the more covert CIA, and the Voice of America became the International Broadcasting Service (IBS) of the USIA.

11.2 Armed forces broadcasting

The need for some kind of broadcasting service to entertain the US forces overseas first manifested itself in 1942. By this time US soldiers were already in action, and defeats in battle had affected morale. Some soldiers took to writing to NBC and CBS requesting suitable programmes be broadcast. Meanwhile, others in the Pacific islands had taken to listening to 'Tokyo Rose', a propaganda broadcast put out by the Japanese. This programme was a high-quality production, offering the best swing music from such orchestras as Boston Pops Philharmonic Orchestra. It was a perfect blend of good entertainment interspersed with intriguing, low-voiced talk by a female announcer who became known to one and all as Tokyo Rose.

Other service units overseas showed a more inventive spirit, by putting together low-powered radio transmitters from junked signal corps components. With these transmitters, and a selection of records, programmes were radiated over the base areas. Some, with greater ingenuity, picked up SW broadcasts from the USA and re-broadcast these programmes.

This spontaneous activity generated the idea for special programmes for forces overseas, but it was a task for which OWI was not suited. A factor that pushed the idea to priority status was that more and more service personnel were listening to enemy broadcasts from German and Japanese sources. The task was turned over to a Hollywood agency Young & Rubican. The mapping of a worldwide forces radio service began. With all the facilities of Hollywood at its disposal, a budget to suit and a galaxy of stars, it began production of 'Command Performance'.

By January 1943, Armed Forces Radio Service was operational with 21 outlets; by the end of that year it had grown to 306 outlets in 47 countries. Each outlet received some 42 hours of recorded material, either flown in from the states or transmitted on short waves for pick up by service units. AFRS was a global service, unequalled by any other nation in the war. Wherever US forces went, AFRS followed. In many instances AFRS broadcasting stations were in action in front lines. In most cases the broadcast transmitters were low power, with a power of 50 W, but some put out up to 400 W or even 1000 W.

The kind of shows broadcast on AFRS were of the finest standards Hollywood could offer. 'Command Performance' meant exactly that. Any US serviceman could write to AFRS Hollywood with a special request, in the confidence that the request would be honoured. No Hollywood star would refuse to take part. So the American GIs listened to the best that Hollywood could dish up: Bing Crosby, Dinah Shore, Judy Garland, Frank Sinatra, the Andrew sisters, Jimmy Duranti, and many more.

This was exactly the kind of entertainment the American forces had wanted, and it went down even better than the swing music of Tokyo radio and its sex-charged Tokyo Rose. By late 1943 to early 1944, AFRS was receiving letters to this effect. Typical of these were letters from an American army nurse in Iceland and a British lieutenant in Egypt:

Command Performance
Hollywood

327 Military Hospital
Iceland

Dear Sirs;
I am writing for a group of army nurses in Iceland to tell you how much we enjoy listening to your broadcasts. It brings a laugh and is almost as good as a letter from home. Our request is: Would you please have Frank Sinatra sing 'You would be so nice to come home to' for all the old maids in quarters 46 in Iceland. We sincerely appreciate all you have done to keep up morale. Our thanks to you for all the time and effort you freely give to make these broadcasts possible. Thanks a lot and we shall be listening for our song.

Judy Garland
c/o Command Performance
Hollywood California

254 Heavy AA Regt RA MEF

Dear Judy;
As you can see we are stationed in Egypt and we can't tell you how much we appreciate your songs. Next time you are on Command Performance what about letting us have 'me and my girl' again — can you let us have a photograph of yourself? We wish you all the luck in the world and hope to hear you soon.

With the task of entertaining US troops overseas detached from OWI operations, and running smoothly under the charge of the AFRS, the OWI was free to concentrate on its original task—propaganda broadcasting to the neutral and enemy nations. But a mutual involvement continued. Obviously, enemy nations and enemy forces could listen to AFRS broadcasts. This invited the thought that items of news designed to confuse and alarm the enemy, could be broadcast on the AFRS network, in other words a variation on the theme of black broadcasting.

11.3 'Over there': US broadcasting in Europe

In the spring of 1944, preparations were being made in Washington and London for what was described by a Hollywood agency as 'the greatest show on earth'. A story in the US magazine *Broadcasting* went even further: 'For all of us alive today, the biggest story since Creation is about to unfold'.

The story in question was the forthcoming invasion of Hitler's Fortress Europe. Whatever else may have been written about this phase of the Second World War, the fact that cannot be denied is that it was largely an American operation. Britain alone could never bring about the defeat of the Wehrmacht; it had had that opportunity in 1940, when it was faced with a numerically inferior foe, yet was routed from France in a matter of weeks.

Now, the Allied armies assembled in southern England waiting for the signal to invade Europe were battle-trained and equipped with superior weapons, made possible by the massive output of weapons and materials from the American industrial heartland. Waiting with the Allied armies were the media. London was the focal point of this activity, and as the days went by more journalists joined the military top brass—and even the exiled royal families of Europe—taking up residence in London. Claridges, the Savoy, and the Dorchester were packed. All the major broadcasting companies of the USA had set up infrastructures for news gathering. The BBC had finalised its arrangements, and the Political Warfare Executive was ready to expand its black broadcasting activities on the Continent.

Among those Americans whose arrival in England had gone unnoticed were Colonels William Paley and David Sarnoff. Both were connected with psychological warfare operations; Colonel Paley was in charge of the Psychological Warfare Department. For four years Britain had conducted its own war against Germany, bombing its cities and saturating Germany and the occupied countries with a massive blanket of propaganda, spreading terror and confusion. Now, the American counterpart, the PWD was set to go into action.

Working in conjunction with the BBC and PWE, the PWD stockpiled ready-made programmes to be broadcast to Germany and the occupied countries on D-Day. These included recordings of speeches made by the prime ministers of governments in exile; Poland, Czechoslovakia, Belgium, Luxembourg and the Netherlands, King Haakon of Norway, and, of course General de Gaulle. All these recordings were now ready for broadcasting all over Europe over the BBC's many transmitters, and those controlled by American broadcasting agencies in Europe.

On 30 April 1944, six days before D-Day, ABSIE put out a historic broadcast to mainland Europe with the words: 'This is the American Broadcasting Station in Europe... In this historic year, 1944, the Allied radio will bring you tremendous news... We shall give you the signal when the hour comes for you to rise up against the enemy and strike.' It was irrelevant that this broadcast was not coming from mainland Europe, or whether there was an effective resistance movement in all the countries that heard the broadcasts (and there were some with no credible resistance); the point was that the Germans could never be sure. It was psychological warfare.

On 6 June 1944, from the early hours of dawn the Allied armies were hurled on the shores of France. With them came the war correspondents, armed with nothing more dangerous than a microphone and suitcase speech recorders. The techniques and equipments used had already been deployed successfully in the Allied landings in southern Italy. So well trained were these war correspondents, who brought live actions from the front lines into people's homes, that few ever realised that these eye-witness reports had been recorded, censored and if necessary changed before being released for broadcasting. War correspondents from NBC, CBS, BBC, ABSIE, AFRS and OWI followed behind the advancing armies like camp followers as the invasion gathered strength.

One of the priorities of the invading armies was to capture intact any radio station or communications centre before the retreating German forces could destroy it. France had many such installations around Normandy, and further inland. In the list of such strategic installations, pride of place was given to Radio Luxembourg, the most powerful radio station in Europe, which had been used so effectively by the Germans against Britain since May 1940. The need to seize Radio Luxembourg intact with all possible speed had been one of the reasons behind Colonel William Paley's arrival in London in April 1944.

The American 12th Army under General Patton had been assigned the task of capturing the massive transmitter and antenna on Juglinster Plain. A special detachment force was created for this purpose. In early September this task force reached the outskirts of Luxembourg, and by 22 September Radio Luxembourg was on the air as an Allied propaganda broadcasting station. It was a most remarkable achievement, and one which has received much less than the credit that is due to those who took part. Events were helped by the station engineers, who with remarkable foresight and optimism had buried a spare set of transmitting tubes in the grounds of the villa.

American Forces Broadcasting had already set up broadcasting stations in Le Havre, Rheims, Biarritz and later Paris, but Radio Luxembourg was the most prized possession. US Signal Corps engineers also found a priceless collection of pre-war gramophone records of American dance bands—Guy Lombardo, Benny Goodman, Dorsey Brothers, Glen Miller—recordings that had been used in German propaganda broadcasts to Britain by William Joyce, alias 'Lord Haw-Haw'.

Psychological warfare strategists from PWD were now in a position to implement fully the plans that had been drawn up six months earlier. From 7 a.m. to midnight, Radio Luxembourg carried on broadcasting as an overtly American-controlled entertainment radio station.

At midnight the transmitter went off the air while Signal Corps engineers made adjustments to its operating frequency. At 2 a.m. it came up on power on a different frequency, at reduced output power; the announcers had changed from American to German. Its new identity was not anti-Nazi, because this would have given the game away. It came on air with the station announcement 'Twelve Twelve calling'*, using 30 kW of power. It carried scrupulously accurate reports of the military situation behind German lines; it used no music of any kind, only a few German voices of correct regional quality to suggest an underground movement within Germany. That the station was winning the confidence of German listeners became obvious when captured German soldiers began to quote the station '12–12'. The trust developed enabled it to become a fearful weapon. The role of '12–12' was to cause chaos, confusion, alarm, uncertainty and fear for what the future held for the German population. By reporting the false presence of tanks in the vicinity of Nuremberg and Ludwigshaven, the station's broadcasts caused panic in the streets. After 127 nights on the air, 12–12 disappeared without trace; it had served its purpose, and Germany's fate was sealed.

Meanwhile Radio Luxembourg continued to broadcast by day to the German people. It featured a daily programme called 'Front Post', which was a sort of daily newspaper using a host of propaganda devices. A million copies of a leaflet headed '12–12' were dropped by air—the first time a leaflet had been named after a radio station.

Radio Luxembourg had its grimmer side. According to reports, two German soldiers were captured nearby, allegedly spying. They were tried as spies, sentenced, taken away and shot. The German listeners heard the click of rifle bolts, the shouted command 'fire' and the reverberation of the shots through the microphone—the first ever on-the-air execution.

The war with Germany came to an end on 8 May 1945, with the unconditional surrender of Germany's armed forces. Yet Radio Luxembourg was not returned to its rightful owners for over another year. During this time, it is believed that Churchill wanted to use it as a propaganda weapon against the Soviet Union.

*Radio station 'Twelve-Twelve' was so-called because it used a frequency of 1212 kHz.

Chapter 12
Japanese wartime broadcasting

As is the case with most government-funded broadcasting services, Nipon Hoso Kyoka, the Japanese Broadcasting Corporation, in 1941 had a national and an international broadcasting service. At that time, before Pearl Harbor, the overseas broadcasting bureau was transmitting international programmes to the world over a network of SW transmitters—some beamed towards Western Europe, Scandinavia and the Soviet Union, others to Canada and the USA. The programmes were produced in English, French, Italian, German and Russian.

At the time, ordinary people accepted that the Japanese attack on the American naval base of Pearl Harbor was one of surprise. However, the attack had been preceded by a fairly lengthy period of strained relations betwen the two countries, and there is considerable evidence suggesting that President Roosevelt knew the attack was coming; one theory is that it was part of his strategy to get America committed to the war. However, the severity of the raid and its overwhelming success were a fearsome shock to American self-confidence.

Pearl Harbor was merely the first of a long series of victories by Japanese Imperial forces. It was soon followed by surprise air attacks on widely separated parts of South-East Asia, stretching from Burma to the Dutch East Indies. After the speedy conquest of Malaya, the British fortress of Singapore, once thought unassailable, fell to the Japanese, when the British garrison of 65 000 regulars surrendered to 15 000 Japanese troops on 10 February 1942. One month later Batavia, capital of the Dutch East Indies, fell to Japanese forces. In May, the Japanese took the Philippines, and with them the American base at Corregidor. A few months later, Hong Kong, the second jewel in the crown of the British Empire, fell without a shot being fired. The sun had set over Britain's Far Eastern colonies. Within a few months, Japan had acquired an empire which extended from Burma through to Java, Sumatra, Bali, the Solomon Islands and New Guinea, and stretched a quarter of the span of the globe, from Burma to the Midway Islands.

The sheer speed of the Japanese onslaught meant that many vital strategic installations were seized intact. These included radio installations for both broadcasting and SW communications. In Hong Kong, the vital wireless installations on Stonecutter's Island were taken in working order; some were put to military use, while others were used to extend the voice of NHK Overseas Broadcasting Bureau.

Figure 12.1 *Wireless station captured by the Japanese army at Stonecutter's Island in the South China Seas. Officers' Geisha House, 1941*

Following its string of victories, Japan had acquired a radio broadcasting network of unequalled proportions. By the skilful use of SW pick-ups from NHK studios in Japan, NHK could beam its broadcasts with even greater power than before. Very soon the voices of Japanese operators dominated the airwaves of South and East Asia. But the evidence is that in the beginning Japan did not use these facilities to their full extent. NHK established some broadcasting studios in Batavia, Manila, Singapore and Hong Kong, but there is little evidence of these being used for propaganda purposes in the first year of the war with America (although broadcasts were beamed to the West coast of America, which had a large percentage of Japanese-Americans).

Equally, the Japanese high command showed little initiative or enthusiasm for becoming involved with propaganda broadcasting; it was beneath the dignity of the Imperial army to engage in such unethical practices, and probably contrary to the spirit of Bushido. The Japanese could afford to adopt such an attitude: they were victorious in battle, had acquired an empire for the emperor and the army had proven itself in battle against American, British and Australian troops.

However, as the pace of the Japanese victories in the field began to diminish, interest was generated in commencing propaganda broadcasting. After some consultation with NHK the number of hours allocated to broadcasting American type programmes was increased. At the same time, it was agreed that a psychological warfare programme should be initiated against American troops holding some territory in the Pacific. A special department was set up for the purpose, co-ordinating the activities and involvements of the foreign office, the home ministry, the Domei news agency and a few lesser agencies. Overall responsibility for propaganda was invested in the Eight Section G2, a department of the general staff headquarters. The leadership of this new department was contested by all branches of the armed forces and the Foreign Office, but in the end it was the Imperial army that came to have the largest say in the matter. Major Shigetsugu Tsuneishe was placed in overall command, and continued to hold this post until Japan's surrender in September 1945.

The department's first attempts at psychological warfare used printed material, so it may be said that they were copying the first attempts by the British government in the use of propaganda leaflets. However, the first magazines dropped on American soldiers were an improvement on the British leaflets that had used ridicule against Hitler. The Japanese propaganda magazines were clearly aimed at lowering the morale of the American GI, using such methods as depicting full-busted American blondes back home pining for their loved ones in the South Pacific.

Although there might have been truth in this simple message, it failed to have much effect on the American soldier. But the Japanese were learning. The power of Allied psychological warfare was beginning to take effect on Germany and the occupied countries of Europe, and there was no reason why it should not be effective in the Pacific, provided a formula could be found.

It was decided to adopt the use of the spoken word, with the aid of radio transmitters using the short waves, and to aim these broadcasts directly at the American troops. In the beginning the problems must have been immense. The project called for a team of producers, script writers, typists, translators and news presenters. But the logistic problems were minor difficulties compared with that of producing scripts in the English language. Because of the radically different syntax of the two languages, the end result could easily prove to be comedy, rather than the intended result which was to be psychologically frightening.

From Eight Section G2 the call went out to find suitably qualified people. Major Tsuneishe is generally credited with the idea of employing American and Allied prisoners of war. Orders were duly sent out to all Japanese theatres of war, and the prisoner-of-war camps in Japan, to screen all POWs for experience of any kind in the related disciplines of scripting and studio presentation.

In a few weeks the perfect candidate had been found and sent to Tokyo G2 headquarters. He was Major Charles Hugh Cousens, the quintessential upper-class English gentleman with a background of broadcasting in Australia. Captured at Singapore, Cousens was then a brigade commander with the Australian army having re-enlisted on the outbreak of war, but he had first been commissioned at Sandhurst into the British army in 1930.

It was his civilian occupation between these two periods of military service that really interested the Japanese authorities. After his emigration to Australia before the war, he became station announcer for radio station 2GB Sydney. This single fact was to change his life for the duration of the war. Cousens was interviewed and questioned by Tsuneishe concerning his experience in broadcasting. He was then invited to co-operate with the Japanese authorities in setting up a broadcasting service to American soldiers. When he refused it seems he was given the straight choice between working for NHK and death, which probably made the decision to co-operate an easy one to make.

Within a month Cousens had been joined by two other POWs. These were Captain Wallace Ince, and Lieutenant Norman Reyes, US Army, aged 20. both had been captured after the fall of Corregidor in the Philippines. As with Cousens, the background of these two soldiers was almost tailor-made to suit the requirements of NHK. Ince had been in charge of a subversive broadcasting radio station inside Corregidor, which had purported to be a freedom station, announcing itself as 'The voice of freedom', and Lt Reyes had been his assistant.

Ince and Reyes were given the same treatment in Tokyo as Cousens, with the result that both agreed to co-operate with NHK in the preparation and presentation of radio broadcasts. Cousens made his first broadcast for the Japanese authorities in October 1942. By all accounts it was less than satisfactory; an Allied monitoring station noted that the broadcast by a POW gave the impression that it was being read from a prepared script.

It was vital to the success of the project that Cousens' co-operation was assured; he was the senior officer amongst the POWs and the one with the most experience of broadcasting. To secure his co-operation, the Japanese authorities promised better living conditions, a clothing allowance and other privileges such as visits to geisha houses.

As the months went by, all three officers were given additional responsibilities including the checking of draft scripts for errors of grammar, syntax and the like, and later were allowed to write their own scripts for broadcasting. After the war had ended, they claimed that this freedom to prepare scripts had enabled them to slip in hidden messsages and sentences with double meanings, that would be obvious to the GIs for whom the propaganda broadcasts were intended, and this may well have helped to undermine the Japanese effort.

In early 1943 Eight Section G2 embarked on a new project, devised in conjunction with NHK and its foreign service monitoring station. The idea was original and well thought out. SW monitoring stations would track and copy broadcasts from the many radio stations in US cities. These local radio stations broadcast news of the kind that the American soldiers in the Pacific were not likely to hear from army sources: local disasters, fires, major accidents, shipping losses etc. These news items would then be recorded, transcribed, edited and retransmitted to US forces over the NHK SW network. Disturbing broadcasts such as these could undermine the morale of soldiers away from home, particularly if the news was true. The Japanese had learned that propaganda based on truth is more effective than any propaganda based on lies, and it was the truth they were telling the Americans.

12.1 'Zero Hour'

In the spring of 1943 Eight Section G2 embarked on a new project: a radio programme called 'Zero Hour'. By this time NHK had developed much experience in planning and scripting high-quality propaganda programmes. The idea in essence was to make American soldiers homesick, tired of killing and waiting to be killed. This motive should be seen in the context of the life of the American soldier on lonely Pacific islands, where life consisted of just that: eating, sleeping, killing, deprived of a normal sex life and the privilege of feminine company. By beaming powerful SW broadcasts to these troops, with programmes that stirred the sensitivity and imagination of the listener, the Japanese authorities stood a good chance of accomplishing their objective. The programmes would remind the soldiers of the good things in life which they were missing out on.

Zero Hour was broadcast at 7.15 in the evening, coinciding with a time of boredom and loneliness after a hurried meal eaten in dismal surroundings, or in foxholes. In June 1943 a US news bulletin reported:

> 'Between the Tokyo broadcasts and the intermittent Japanese air raids life is far from dull. Tokyo has beamed a short wave broadcast to the Russell Islands and Guadalcanal. The programme is far from dull and is called "zero hour". The fellows like it because it cries over them, and really feels sorry for them. It talks of food, the girls back home they miss, and how the munition workers back home in their town are stealing their girls and wives.'

The tragedy was in its truth. The GIs, the Japanese authorities and the American military authorities all knew it to be true.

By the summer of 1943 Japanese propaganda broadcasting was proving to be effective. Worse still, from the American viewpoint, it was capturing larger listening audiences. In the autumn, Major Tsuneishi decided to capitalise on its success by giving the broadcasts an all-American radio station flavour. The propaganda base was to be expanded by bringing in more POWs to work for NHK.

Fifty-three allied prisoners were brought in. These were housed in a facility in the Kanda district of Tokyo, which was given the name 'Surnagi Technical Research Centre'. To the first POWs who arrived it was known as Bunka camp. From December 1943 to February 1944 POWs arrived at Bunka camp. All had been screened for their experience and suitability for the work, and given the straightforward choice of co-operating or dying. They co-operated.

Some were put to work in the receiver monitoring stations operated by the army, the air force, the navy and the Domei press monitoring station, while others joined Cousens, Ince and Reyes in the studio work. Lieutenant Edwin Kaebfleish, Ensign George Henshaw, and Sergeant John Provo were selected as announcers for a new programme, modelled upon 'Zero Hour' but with new script writers. The programme was called 'Red Sun'. Other prisoners were put to work on other programmes; 'War on War', 'Post-War Calls', 'Australian Programme' and 'Civilian Air Programme'.

The tall, long-limbed American POWs were soon popular with the female typing and secretarial staff of NHK overseas broadcasting bureau. Associations and romances developed from working in proximity. Lt Reyes, who was single, was permitted to marry Katherine Morooka, one of the Japanese announcers.

12.2 Tokyo Rose

Major Cousens had become friendly with Iva Togura, a Japanese-American girl who, stranded and unable to return to California, had found some suitable typing work with NHK. Iva was pro-American. Cousens selected Iva to work as an announcer on his radio show, though meeting with a little opposition initially from his Japanese superiors. Cousens knew what he was doing: he wanted a genuine American female voice to join the team, which now comprised eight in all.

Cousens said later at the trial that he had chosen her because of her attractive voice with a touch of WAC officer quality about it. Iva Togura accepted the job at a rate of 100 Yen per month. By so doing she had taken the first step along the road to being branded as a traitor.

By mid-1944 NHK overseas broadcasting bureau had honed and crafted its skills in making authentic American-style radio programmes. By patience and dedication the Japanese had overcome all initial problems and were setting a new standard in radio propaganda, which was unequalled by anything existing. Although not modelled on the broadcasting stations of the German Afrika Korps, which broadcast Lilli Marlene to both German and British forces, it had achieved the same kind of universal popularity.

'Zero Hour' was as good as anything that came out of America—it was well produced and directed. A typical hour's programme included the following features:

- Loud opening number: 'Strike Up the Band', by the Boston Pops Orchestra
- Messages from POWs: 'Hi mum, this is Corporal X, we are OK but we need socks and food'
- The Orphan Annie Show: 20 minutes of high-quality American jazz and semi-classical records introduced by Annie
- American home-front news: titbits picked up from America by Japanese monitoring stations
- 'Juke Box': 15–20 minutes of popular jazz
- 'Ted's news highlight tonight': more news from overseas SW broadcasts
- News summary: sometimes read by Charles Yoshii, the Japanese 'Lord Haw Haw'
- Military marches

These programmes were said to have been skilfully put together by Cousens, taking advantage of slang, jokes and puns to get across his hidden message: that 'Zero Hour' was really coming from one American soldier to another. The programme originated in NHK studios in Japan, and was beamed to different theatres of war over NHK SW transmitters.

The official view of the US government was that these programmes were propaganda. This view was not always shared by its armed forces: reports from the US Navy in the Pacific thought that the Japanese broadcasts did a lot for the American soldier on the Pacific atolls. This view was certainly shared by the US Army and the enlisted men who huddled around at the time of sunset to hear their favourite announcer. When the female Japanese announcers did not reveal their name they were quickly dubbed by the American troops: Tokyo Rose, Orphan Annie, Manila Rose, The Nightingale of Nanking and others became talked of as actual persons. There can be no doubt that many a soldier or marine fell in love with a woman they had never seen, nor were ever likely to. They did not know the age of the announcers, nor what they looked like. None of this mattered to the love-lorn soldier, who created in his own mind the picture of the female form.

Of all the female announcers of NHK overseas broadcasting bureau, the one that stood out from the rest was Tokyo Rose. Some names evoke beautiful thought, and the word Rose has few equals. Tokyo Rose became known to all Americans in the South Pacific. But did such a person exist?

Thousands of GIs claimed to have heard her broadcasts, and many said she had actually introduced herself to listeners by that name. Thousands more could remember details of her broadcasts. Her voice had become a sexually charged symbol of someone far more beautiful than any of her rivals from Hollywood, from Betty Grable down. Legends have a way of becoming more powerful than truth, and it would seem this was the case with Tokyo Rose.

A book entitled 'Tokyo Rose' has tried to trace the history of this siren of the South Pacific. The American FBIS closely monitored Japanese propaganda broadcasts, and many recordings were taken for archival purposes; yet not one of these has the name Tokyo Rose, nor is the name found in NHK broadcasts. There are, however, some recordings of an announcer calling herself Orphan Annie. The word 'orphan' was often used in Japanese broadcasts to describe the fate of the Australian forces, caught up in a war not of their making, sent to defend Singapore, captured and abandoned to their fate by the British. NHK itself has no record of any person by that name in its employ. NHK is said to have first heard of Tokyo Rose in a report from, of all places, neutral Sweden during the war. But legends die hard, and indeed in the case of Tokyo Rose they gathered more momentum as the Pacific War progressed to its conclusion.

From 1944 the US Government took some steps to combat the effects of Japanese propaganda programmes by setting up radio shows of a comparable quality and style. 'Voice of America' SW transmitters beamed such programmes to its troops in the Pacific zone. The highly successful 'Command Performance', scripted and produced by Hollywood specialists, employed a galaxy of stars to entertain the troops.

As America's sheer military, economic and industrial might began to tell, the US forces slowly rolled the Japanese back across the Pacific. As they gained ground, the US forces were able to fly into the recaptured islands stars from Hollywood to entertain the marines. This service was co-ordinated by the Armed Forces Radio Service, which had taken over responsibility from Voice of America to combat Japanese propaganda broadcasting.

By the end of 1944 America was firmly dictating the war. Japan's defeat was only a matter of time. From this point on, the Japanese propaganda output from NHK underwent a change in character: the propaganda content increased, while the entertainment content began to lose a little of its magic. And yet, for many of the veteran US marines who had crouched in their foxholes to escape the nightly bombing from Japanese planes, Tokyo Rose was still the one they could never forget, the one who had cried over them. Any one of a dozen female announcers could have been the one they listened to; Ruth Hayaka, June Suyama, Margaret Yaeko Kato, Katherine Fugiwara, Katherine Morooka, Mieko Furova, Mary Ishiti, Iva Togura and Myrtle Lipton, who broadcast from the Manila studio. Yet the legend of the real Tokyo Rose lived on.

After the war the US government embarked upon a witch hunt to find the siren of the Pacific. In June 1949 Iva Togura was indicted on a charge of treason against the United States.

Typical 'Orphan Annie' broadcast, 14 August 1944:

> 'Hello you fighting orphans of the Pacific. How's tricks? This is after her weekend off, Annie is back on the air, strictly under union hours. Reception OK? Well it better be, because this is all request night and

> I've got a pretty nice programme for all my favourite little family, the wandering boneheads of the Pacific Islands. The first request is for none other than the boss, and guess what?—he wants Bonnie Baker and "My resistance is low". My, what taste you have sir, she said...
>
> 'This is your little playmate Orphan Annie, and by the way, wasn't that a lousy programme we had last night? It was almost bad enough to be the BBC, or its little sister ABC
>
> [Following the news section, read by Ince] Thank you, thank you, thank you. Now let's have some real listening music—you can have your swing when I turn you over to Zero Hour. Right now, my little orphans, do what mama tells you. Listen to this—Fritz Kreisler playing 'Indian love call'...Boy oh boy, it stirs your memories, doesn't it? Or haven't you boneheads any memories to stir? You have! well here's music 'In a Persian Market' played especially for you by the Boston Pops Orchestra...Orphan to orphan—over.'

These broadcasts are believed to be scripted by Major Cousens, who coached Iva Togura in reading technique. The scripted words 'thank you, thank you, thank you' are a deliberate 'wipe out' device to clear the listener's mind of what has gone before.

Chapter 13

Treason by radio

History itself is the best judge of the importance that was attached by adversaries of World War II to propaganda broadcasting as a weapon of psychological warfare. Perhaps the supreme example is supplied by the way in which the Allied victors sought out those who had been guilty of working as programme presenters, tried them for treason and passed sentences that can only be described as barbaric.

13.1 'Lord Haw-Haw': William Joyce

William Joyce was an upper class Englishman who had joined the German broadcasting authority Reich Rundfunk in the August of 1939 in the capacity of station announcer, whereupon he was posted to the Hamburg studios as English announcer—a civilian occupation. Joyce's work was of a high standard and he became well liked by the head of department.

At different times Joyce's voice could be heard over many different radio transmitters, although he himself worked at the studios in the Charlottenburg palace in Berlin. At other times he broadcast from the Bremen studios, and on a few occasions he broadcast from the basement studios of Radio Luxembourg.

By using powerful radio transmitters the Germans could project a powerful signal to England. These transmitters included a 120 kW installation at Osterloog, near Norden, built specifically to broadcast to England, and the more powerful Radio Luxembourg, which was seized intact by the German army in May 1940.

Joyce, in his broadcasts to listeners in England, used truth as a powerful weapon, which is why the British government waged a vendetta against him by depicting him as a buffoon, and creating the impression in the minds of the British people that he was a traitor. So successful was this campaign by the media that few people in Britain thought otherwise.

All governments have a fear of propaganda directed against them, true or false. Following the invasion of Fortress Europe, the 12th Army Group was given two priority tasks: the seizure intact of Radio Luxembourg, and the capture alive of William Joyce. The transmitter station at Luxembourg was seized 11 weeks after D-Day. Joyce was working at the Hamburg studio, with the result that his capture did not take place until May 1945, when Germany surrendered.

Joyce was held in custody by British forces in Hamburg for a special reason: the British government needed time to amend the Treason Act to include those of Irish descent. This change took place after Joyce was in custody, so Joyce's defence was that at the time of committing the offence, he was a neutral subject.

Nonetheless, Joyce was charged with high treason. After a three day trial—possibly the fastest treason trial in modern British history—he was found guilty and sentenced to death by hanging. His appeal was dismissed outright, and he was eventually hanged at Wandsworth Jail on 3 January 1946. It was a verdict that should have aroused some considerable public unrest. In plain simple terms, he was hanged for no greater crime than having worked at a civilian radio station.

13.2 Tokyo Rose

Meanwhile, the US State Department investigated several cases and put on trial some of its citizens. Governments in general have a fondness for the word 'traitor'. By the process of seeking out, trying, and punishing those found guilty a nation elevates its moral status in the eyes of its citizens—or at least that is what governments believe.

There are times in a nation's history when it is politically expedient to take such actions in the name of unity. McCarthyism in America was a means of unifying the nation against the threat of Communism. The years that followed the end of the war with Japan were another such period. Memories of the war were still vivid in many American citizens, and it was convenient to have a few treason investigations.

The US Government indicted five people alleged to have co-operated with the enemy by broadcasting over enemy radio transmitters: Robert Best, Douglas Chandler, Martin Monti, Herbert Burman and Mildred Gillars were accused of working for the enemy. In some cases the evidence against the accused was overwhelming. However, in at least one case there was a miscarriage of justice. This occurred in the trial of Iva Togura, convicted of having been the elusive, enigmatic Tokyo Rose.

Togura was a US national of Japanese extraction, who had gone to Japan some time before 1941. She was one of many US citizens trapped in Japan, where eventually she got a job as a typist working for NHK. On account of her clear American accent, she was eventually persuaded to work in the scripting department of NHK overseas broadcast service.

After the Japanese surrender in September 1945, the city of Tokyo became a Mecca for the media. Journalists and cameramen sought permission from the Allied military government to visit the city in the hope of finding exclusive stories. Among these were two freelance journalists who had come with the intention of finding the elusive Tokyo Rose.

Eventually, after gaining access to NHK files, the name of Iva Togura was pulled out, along with some others. With the promise of a handsome payment for a story, Togura agreed to co-operate. The outcome was predictable: stories appeared in American newspapers claiming to know the identity of Tokyo Rose. In 1946 the Allied military government in Tokyo brought Iva Togura in for questioning over alleged co-operation with the enemy; subsequently she was released, it being agreed there was insufficient evidence against her. But the press did not give up so easily. Further articles appeared about Tokyo Rose, and it was even claimed that Togura had given away autographed pictures of herself—she was learning to be a star, and soon she would learn what it was like to be a traitor.

Over the next three years the memory of Tokyo Rose was kept alive by the American press. Whether it was due to this, or because the State Department wanted to impress the people, we shall never know, but in 1949 Iva Togura was taken into custody and formally charged with treason.

The case for the defence was that there was no such person as Tokyo Rose, and it produced evidence to that effect. It would seem that the name Tokyo Rose had been dubbed to any female announcer in the employ of NHK. The Japanese authorities had no employee by that name, and the first the Japanese broadcasting authorities knew of her came from a news report in a Swedish newspaper during the war. Iva Togura did not deny that she had worked for NHK, but that she had always acted under orders from her superiors, who included Allied officer prisoners of war, some of whom testified in her defence.

Nevertheless, Iva Togura was found guilty in October 1949 and was sentenced to ten years' imprisonment. The trial had lasted 13 weeks. Iva Togura served the ten years for a crime she was not guilty of, and in 1976 was granted a full and unconditional pardon. In retrospect it seems likely that it was politically prudent and convenient to find and punish those alleged to have co-operated with the enemy, in the same way that 20 years on it was politically expedient to bury the hatchet with Japan, which had become America's strongest bulwark against communism in the Far East.

Another famous trial was that of Mildred Gillars, alias Axis Sally. Unlike Tokyo Rose, who could have been any one of a dozen female announcers used by NHK, there was only one Axis Sally, the name given to Mildred Gillars. She broadcast from Radio Berlin over SW transmitters situated at Zeesen, just outside Berlin. Gillars was a native of Maine, USA who became a disc jockey for Radio Berlin in the mid 1930s. During the war years she was promoted to propaganda broadcaster, a job to which she was well suited. Gillars specialised in antisemitic propaganda, which did not go down well in America, where Jews were (and still are) a powerful and organised lobby. However, what really damned her was her broadcasts to America a few weeks before D-Day. She painted a dramatic scenario in words of drowning American GIs whose ships were attacked by German U boats; though bloodcurdling, it was for many an unfortunate soldier and sailor the truth.

Mildred Gillars continued to broadcast for Reich Rundfunk for the duration of the war, but she was a marked woman in the eyes of America. In her defence at the treason trial it was said she was the lover of Max Otto Koischewitz, her head of department, a Svengali-like figure who had turned her into a puppet. This availed her little; she was found guilty of treason and sentenced to 30 years. If there is a moral in this case for would-be propaganda broadcasters working for an enemy it is to steer clear of broadcasting the truth.

The case against Gillars was considered complete because she freely admitted to having broadcast for the Third Reich, and in the eyes of many US citizens her crime amounted to worse than murder because she attempted to demoralise US troops before battle. As far as the Americans were concerned she was as infamous as Lord Haw-Haw who had broadcast to the British people. There was also a resemblance in the way both had been dubbed with names intended to evoke feelings of hatred and disgust.

Another famous propaganda broadcaster was 'Manila Rose'. This was the name given to a female announcer who broadcast for the Japanese over a SW circuit beamed from Manila (occupied by the Japanese) to American forces in the New Guinea area. There is no doubt that Manila Rose was popular with the GIs: as with Tokyo Rose, the name engendered a feeling of love and tenderness rather than hatred and loathing.

The credit for the affection shown and expressed by American forces towards these female announcers of NHK must go to the Japanese broadcasting authorities who, in the face of huge language and cultural barriers, succeeded in putting together some very high quality entertainment programmes that any American radio station would have been proud of. Although it cannot be denied the underlying motive was to make American soldiers sick and tired of senseless and bloody war, it was skilfully done. And, although not recognised at the time, it was an example of the Japanese dedication to quality that would later enable Japan to dominate the West.

Manila Rose's real name was Myrtle Lipton. Her programme, called 'Melody Lane', was broadcast between 5.30 and 6.30 p.m. Manila time. By all accounts it had a certain endearing charm about it, modelled perhaps on Zero Hour. Myrtle Lipton was never brought to face a trial for treason in the American law courts; nor indeed was she ever taken into custody as were the others. It was said that she was irresistably charming to men and it is rumoured that she disappeared with the US colonel who had gone to question her. Nothing has been heard since.

Chapter 14

Woofferton SW station

Woofferton radio station, Herefordshire, England, today has been eclipsed in size by a number of SW stations in Europe: Rampisham in England, Issoudun in France and Wertachtal in Germany. Nevertheless, its different roles during World War II, its role during the Berlin blockade and the subsequent Cold War entitle it to a place in the history of SW propaganda broadcasting.

Its origins date back to 1942, a dark period of Britain's history when it was struggling in every theatre of war and the possibility of total defeat loomed high. At this time the Government approved budgets for the greatest expansion ever in Britain's propaganda broadcasting capability: the building of a new generation of high-power SW transmitting stations. The plan called for the building of three powerful radio station sites: Skelton A and B in Cumberland, Rampisham in Dorset, and Woofferton in Herefordshire. This grand expansion scheme was designed to increase the BBC's external services to a total of 37 high-power SW transmitters by 1943, in contrast to the network that existed in 1942—eight medium-power transmitters at Daventry of a pre-war, now aging, design.

During the war years, the BBC stations of the external services were identified by numbers rather than place names as a security measure. Woofferton was named OSE-10 and was built on a 3 km^2 site, 75 m above sea-level, 8 km south of the market town of Ludlow, and 11 km north of Leominster. The design of OSE-10 differed quite radically from all previous stations in that it was based on American technology. There were two chief reasons for this; the unavailability of British-built transmitters, and the more modern design of American-built transmitters, which, unlike their British counterparts, did not need a building with a crypt to house the valve cooling plant. The high water table at the Woofferton site would have made it almost impossible to construct a suitable building.

By January 1943 the first of the six RCA 50 kW SW transmitters arrived in England; the remaining five sets did not reach Woofferton until July that year. The station eventually came on the air on 17 October 1943. These 50 kW transmitters embodied everything that was good in American transmitter technology: high-level modulation using a Class B modulator design, easy accessibility of all major components and the ability to undergo a valve change in a matter of minutes. A unique feature at that time was the use of rubber hose wound in the shape of an RF choke inductance, through which the water coolant

Figure 14.1 *The Woofferton VOA relay station, England. 1944 RCA transmitters and programme input racks. Note the old-fashioned 'dalek' monitor loudspeakers*

flowed to the copper anodes of the power-amplifier triode valves. In common with all other transmitter designs of that era, changing the operating frequency was a manual operation that could take 15 minutes, but even this was a big improvement over the Marconi SWB-18, although the latter was of a more robust design.

The Woofferton transmitter building was itself unusual in that it incorporated

Figure 14.2 *RCA transmitters at Woofferton from 1943 to 1970*

Figure 14.3 *RF drive equipment at Woofferton being checked against the Droitwich transmitter*

a special lightweight roof designed to give way in the event of bombing. Experience gained from the bombing of London had shown that most of the damage resulted from heavy roofs falling in. Short of building bomb-proof roofs, the alternative was to have virtually no roof.

The SW antenna system for OSE-10 consisted of 26 separate dipole arrays, each capable of covering a single broadcast band within the HF spectrum. These were suspended from 15 stayed lattice-type masts varying in height from 46 to 100 m. The complete antenna system was designed for remote selection from the control desk. From this position, any transmitter could be connected to any one of the 26 curtain arrays. The RF switches to facilitate this operation had to be rated to carry 50 kW carrier power. These were mounted on a six-level tower from which the 600 Ω feeders ran to the curtain arrays.

Figure 14.4 *The control position at the Woofferton VOA relay station in the 1960s. Note the old-fashioned jackfields and the old-type drop-flap telephone indicators*

OSE-10 was equipped with its own standby power generation plant for emergency use. Under normal operation the station drew its supply from the public electricity supply. In case of failure for any reason (such as enemy bombing), three supercharged Harland and Wolff 750 hp marine diesel plants could provide a total capacity of 1.5 MVA—more than enough to run all six transmitters at their full rated power for which the normal demand was 900 kVA.

When OSE-10 became operational in October 1943, it was the most modern high-power SW station in the Western hemisphere; but its glory was shortlived, for less than a year later it suffered a major crisis and was eventually closed down for the rest of the war. What happened to OSE-10 has remained secret for 40 years. This period of OSE-10 has not been covered by any historical account of British broadcasting, and even stranger, the BBC's Written Archives Section contains nothing about the year 1944–5 for the station. The discovery of the missing history of OSE-10 came about by pure chance in the summer of 1986.

In late August 1944, I was part of a technical team from RAF West Drayton, which was hurriedly dispatched to the BBC station at Woofferton to dismantle the six high-power transmitters with care, transport them to Crowborough and reinstall them. The high priority to the task became apparent later—I and others had been hurriedly brought back from the Azores a few days before to prepare for this new task. We arrived at OSE-10 on 28 August, the date the station closed.

Working without sleep, we dismantled the transmitting equipment. By 19 September two of the RCA transmitters had been reinstalled at Crowborough,

Figure 14.5 *The 1943 design of six-level aerial switching tower at Woofferton*

Sussex on a concrete base even while a brick housing was being built over them.

These installations and many others much less powerful were a desperate attempt to counter the V2 rockets, which the Germans were about to launch against England. British intelligence had deduced that these weapons were launched by a radio control signal within the frequency range 30–60 MHz, and if these launching signals could be jammed, it might be possible to upset or divert the rocket from its intended trajectory. (The V2 rocket sites in Holland and Belgium were later overrun and put out of ation by the British 2nd Army, but not before at least one had landed in Chiswick, London. To prevent public alarm and possibly panic, the incident was reported by the MOI news to the British press as a gas main explosion.)

This episode of the war, and the involvement with the Woofferton exercise from beginning to end, have stayed in my memory, especially the part where the same team, before the end of the war, was given the task of reinstalling the RCA transmitters at Woofferton. This was a far more difficult task, due to the damage the transmitters had suffered in the hands of the RAF team and the staff at HM Government Communications Centre (HMGCC) at Crowborough.

In 1986, hoping to collect research material on OSE-10, I contacted the Written Archives Section of the BBC. It could find no information on the history of Woofferton during the war, except for a one-paragraph piece in a magazine. Incredible though it may now seem, this episode in the history of the BBC had been erased.

There the matter might have ended but for the fact that a BBC engineer, working at Woofferton since 1959, intrigued by the wartime happenings at OSE-10 and local gossip concerning the mysterious team that arrived to take away the radio station, decided to pursue his own research. It was after his contacting the BBC Written Archives Section that he was eventually put in contact with me.

This researcher was able to piece together the missing chapter from OSE-10's history. Visits to the Public Records Office revealed information in Air Ministry files AVIA 7 2558 and leave no doubt that the BBC was then the instrument of the Ministry of Information, since the RCA transmitters are mentioned as being the property of that department.

There is no evidence to support the theory of British intelligence that the V2 rockets were launched on a radio beam: in fact the V2 was a rocket with a free-fall characteristic on its downward path trajectory. Nor can the expense of the Woofferton exercise and others similar be justified, especially the manufacture by Marconi of many lower-powered transmitters that were deployed around England as a radio countermeasure.

Thus the Woofferton incident was a failure and it became deleted from the history of WW-II. Success is said to have a thousand fathers, but failure is always the orphan. Although all six RCA transmitters were eventually back in service at Woofferton, the station was never the same again: several of the RCA transmitters, the most beautiful radio transmitter of that era, had been badly scarred; damaged front panels and crudely repaired components did nothing to instil pride in the staff that had returned to Woofferton for the first time since the RCA transmitters were taken away in trucks, dismantled into hundreds of pieces.

From 1945 to 1948, Woofferton functioned as an overseas extension, but in June 1948, the station was again closed down, this time for reasons of cost. Britain had won the war at the expense of virtual bankruptcy with a massive war debt. One of the items still to be paid for was the six RCA transmitters, one or two of which had been reduced to little more than scrap value.

It was another war that saved OSE-10. In the aftermath of the communist takeover in liberated Czechoslovakia, and the announcement of the Marshall plan, which supplied aid to those countries that renounced communism, there quickly developed the Cold War, followed later by the Berlin blockade.

On 18 July 1948 Woofferton came back on the air with all six transmitters once again working at full power, like a phoenix risen from the ashes. This time it was being completely financed by the US Office of War Information as part of the Voice of America broadcasting network. Five of the six senders carried two VOA programmes, Blue Stars and Grey Stars, on a single-shift basis between 2 and 10 p.m. daily. VOA programmes from its studios in Washington were relayed across the Atlantic by short wave, picked up at BBC Tatsfield receiving

Figure 14.6 *The original RCA transmitter control unit at Woofferton. Programme keys modulation meters in the centre of the desk with General Electric noise distortion and measuring equipment behind. Note the gloves for wave-changing on top of the desk*

station and then relayed over Post Office land lines to Woofferton.

During this period OSE-10 had been carrying propaganda broadcasts in support of the breaking of the Berlin blockade. The blockade finally came to an end May 1949. The Cold War did not, damage to East–West relations was beyond repair, and on 11 June 1950 Woofferton went into three-shift, round-the-clock broadcasting. All transmitters at Woofferton and other BBC transmitter stations had their outputs pushed up to maximum limits, marking the commencement of barrage broadcasting.

In 1963, as part of the continuous postwar expansion of Britain's propaganda broadcasting service, the BBC began to replace the aging 50 kW RCA transmitters with a more modern design of greater power output. The RCA sets by this time were almost 25 years old, and some significant improvements had taken place in transmitters since the war. Four of the RCA sets were removed to make way for six 250 kW sets of Marconi design (type BD272). Though having five times the power, these new transmitters were physically smaller than the RCA 50 kW, and nowhere near as impressive—the RCA sets were some 9 m in length, which gives some idea of the problems that were involved in transporting these to the HMGCC site in Crowborough during the war.

With the replacement of the 50 kW transmitters, a new power feed had to be brought in from the nearest electricity substation with a step down to 11 kV from 33 kV. Other changes to OSE-10 included an expansion of the site to 5 km^2 to provide required space for additional arrays. The old antenna switching tower, the classic monument to outdated technology, was demolished and replaced with a more flexible trunk-and-spur system. Arrays were now connected to feeders via electrically controlled, pneumatically operated switches on a cross-point matrix system. The number of curtain arrays was increaed to 35, thus greatly increasing the flexibility of target zones. The old standby diesel alternator plant was no longer sufficient to run the entire station, but was retained to provide partial cover.

In 1979 OSE-10 underwent yet another major refit, involving the taking out of service of the last two 50 kW transmitters of wartime vintage and replacing them with four more transmitters, this time the more modern Marconi B6124. The major difference was a much faster frequency changing capability.

Three major station refits in less than 40 years, accompanied by a dramatic increase in the total output power of the station, gives some idea of the importance attached to Britain's voice on the short waves. By way of contrast to its national broadcasting capability, it may be noted that Brookmans Park was constructed in the mid 1930s and did not receive a station upgrading until 1984.

Figure 14.7 *The Woofferton VOA relay station with four additional 300 kW B6124 short-wave transmitters added in 1987*

Chapter 15

The Cold War

The term 'cold war' is not new to this century; it had its origins in possibly the longest-running war in history, a war that has seen many phases from cold to hot. I refer to the war between Christianity and Islam. The term 'cold war' was coined by Don Juan Manuel, a 14th century Spanish writer and historian, who said: 'Cold war brings neither peace nor honour to those who make it'. He might as well have been writing about the Cold War of the 20th century, in which propaganda broadcasting played the key role, equating in importance with the role of the nuclear weapon in a hot war.

The Cold War of our century falls into five distinct phases chracterised by the prevailing relations between the two main antagonists: The United States of America and the Union of Soviet Socialist Republics.

Phase one: the first Cold War, 1946–53
Phase two: oscillatory antagonism, 1953–1969
Phase three: detente, 1969–1979
Phase four: the second Cold War, 1979–1986
Phase five: perestroika, 1986 onwards

Although it is hard to define when the Cold War began, for our purposes the most significant event is when the BBC was instructed to commence broadcasting to the USSR. This was 24 March 1946, and Voice of America followed in February 1947. Of course, neither America nor Britain broadcast to the people of the USSR throughout the Second World War, when it could have been of comfort to the Soviet people — in contradiction to the BBC's official stance, that it broadcasts to other nations for the purpose of promoting a better relationship and understanding. In fact, the BBC suspended its broadcasts to the USSR during the war only after it became an ally.

The seeds of the Cold War were sown before the end of the Second World War when both sides were taking stock of what the future held. Substantial conflicts occurred in 1946, and it was Churchill himself who first coined the expression 'iron curtain', which he spoke of as 'running from the Baltic to the Adriatic'. There is some evidence that in the closing months of the war with Germany, Patton wanted to keep on driving his tanks until they reached Moscow — a view with which Churchill is thought to have sympathised.

When the Truman administration pronounced its willingness to organise anti-communist forces in Greece and Turkey, it aimed to destroy communism and revive capitalism. The Marshall Plan of 1948 openly promised aid to any country wishing to renounce communism. By then communism and capitalism were in direct confrontation: the first phase of the Cold War had commenced.

It began with a particularly intensive propaganda campaign in which both sides sought to maximise the denigration of the other. Western propaganda broadcasts from the BBC and VOA berated the USSR for its totalitarian qualities, and likened the nature of communism to the Nazi regime of the Third Reich on the grounds that both were of a warlike repressive nature. The West was following the theme depicted on the first propaganda leaflets dropped on Nazi Germany in October 1939, showing Hitler and Stalin side by side in a cartoon picture.

Soviet propaganda, coming out of transmitters in the USSR and East Germany, used Marxist philosophy to attack features inherent in capitalism—exploitation of labour and the war-mongering nature of imperialism. It pointed to those parts of the world where Britain's colonial empire was struggling for the right to self-government.

1947 saw the forcible takeover of power from communists, in Greece and parts of Eastern Europe, a result of the Marshall plan. On 6 June 1948 the Soviets closed all roads to the city of Berlin, sealing off the three western sectors of the city. The Berlin blockade had begun. Over the next 11 months the US Air Force flew 180 707 flights into Berlin with supplies. On 20 May 1949 the Soviets lifted the blockade, and conceded a victory to America.

Different reasons have been put forward to account for the origins of the Cold War. Whatever one's theory, the fact remains that the existence of the atomic bomb and the means of delivery, supported by a massive barrage of propaganda broadcasting, made the world unsafe.

The ability of propaganda broadcasting to destabilise a government should not be under-estimated: propaganda from BBC transmitters is generally credited with the overthrow of the Shah of Persia in 1940. It was no coincidence that by the time the war was won in 1945, Great Britain had built up the largest network in the world with 46 high-power transmitters in England alone. This broadcasting network of the BBC external services was augmented by a second broadcasting network operated by HMGCC, including the giant 600 kW medium wave transmitter at Crowborough, Sussex that had created havoc in Nazi-occupied Europe.

By 1946 its massive war debts had brought Great Britain to the point of bankruptcy, yet it possessed the capability to transmit 850 hours of propaganda per week in 45 different languages, then unrivalled by any other country in the world. One of its stations, Woofferton, was mothballed in 1946 as an economy measure, then reopened to broadcast to Eastern Europe. The barrage of propaganda broadcasting by the BBC and HMGCC reached a peak during the Berlin blockade.

The Soviet Union countered this by setting up jamming radio stations to prevent the broadcasts from being heard in Eastern European countries—effectively, censoring the propaganda. The war escalated when the BBC further increased the level of propaganda by curtailing, and later suspending, broadcasting to friendly nations thus releasing more transmitters for the Eastern service. In the words of the BBC, 'It became more of a question of letting the ordinary people hear the western point of view'. In other words, it was attempting to go over the head of an established government and appeal directly to its peoples.

When the Soviet Union increased the numbers of jammer transmitters, the BBC responded by resorting to the use of 'burn in' technique, a method devised in wartime of reading messages in very short sentences, delivered at a slow speed, and broken up by long pauses of silence. This is a psychological technique to permit the mind of the listener to grasp the all-important context.

Broadcasting, counter-broadcasting, and counter-counter broadcasting may be described as ascending states of oscillatory antagonism where the object of the aggressor is to destabilise a government by appealing over its head to the people. The obvious risk of this process is that it will, in the end, provoke an open war. Wars arise (and the Third World War, which almost happened during the 1960s, was no exception) not so much from any nation wanting to go to war, but rather when a nation is forced into a situation from which there is no retreat. Propaganda broadcasting can create such a situation when it is fuelled by the correct political ingredients.

The BBC's line is that broadcasting to another nation creates a better understanding and can therefore defuse an international situation that is potentially dangerous. However, history shows quite a different pattern. In the year up to September 1939, when the British Government was pursuing a policy of appeasement with Germany, the BBC remained silent: no broadcasts in German were made during this period. Similarly apart from one broadcast to the Soviet Union in 1941, the BBC did not broadcast to the Soviet Union again until 1946—the commencement of a policy of confrontation between East and West—and the quantity of these broadcasts increased to suit the tempo of the Cold War.

When governments wish to mediate, negotiations are conducted through duly appointed mediators. Hess flew to negotiate with the Duke of Montrose in 1941 in secret. Negotiations for the surrender of Germany in May 1945 were conducted in secret through a neutral country. In the case of the Berlin blockade, when a very dangerous international situation was wound down in the later years, the BBC wound down its broadcasts to Eastern Europe.

The period of the Cold War that began in 1946 and intensified, also brought to the art of international broadcasting a new image. During the war years the powerful radio transmitter had a covert existence; the locations of transmitters remained as secret as the nature of the use to which they were applied—the broadcasting of subversive and misleading information to confuse the enemy.

In the Cold War there was a reversal of roles: the powerful radio transmitter became the major weapon of war and the real weapons of war were kept on a leash. Thus, from 1946, the art of propaganda broadcasting began to acquire a more overt status. This change in status was responsible for much of the growth rate that has characterised international broadcasting since the 1950s.

Much of the credit for the transition must go to America and to VOA, the international broadcasting arm of the United States Information Agency (USIA). USIA was formed in 1953 and took over the wartime operations of OWI, which had included responsibility for VOA operations. With a greatly increased budget, consistent with America's self-cast role as guardian of the free world, it began to plan for a worldwide broadcasting network, based on the same expansionist plan as that laid down for its armed forces. This broadcasting network is described in more detail in a later chapter.

By the late 1950s 31 nations were engaged in international broadcasting, and the numbers were growing with each year. It should be stressed that not all propaganda broadcasting is aimed at subversion or confrontation. The principal role today of international broadcasting is to act as an instrument for, and a servant of, a nation's foreign policy, and to that end there is almost no limit to the way it can best serve these interests. There are many countries that have no territorial ambitions of expansion; others, such as France, have made a graceful transition from being a nation with an empire, to a nation content to use international broadcasting as a means of projecting its culture to its erstwhile colonies.

A good guide to the purpose to which international broadcasting is put is by an examination of the languages that are broadcast, and the hours allocated to a particular country. During the Berlin blockade almost the entire output of the BBC external services was directed to Soviet Bloc countries. Although this balance has since been redressed, the latest figures for the BBC World Service show that BBC broadcasts to Eastern Europe still outnumber French language programmes by a considerable factor; even the little state of Bulgaria has more programmes directed to it than does France, a world power of some stature. Poland and Czechoslovakia each attract twice as many hours as France, whilst broadcasts in Russian outnumber French by 4:1.

Chapter 16

The Voice of America

In 1953 Voice of America became the international broadcasting service of the USIA. In that year the USIA began to make some long-term plans for its 'voice', which were unique to VOA. It was to be the world's first truly global broadcasting network. Until now, all other countries had located their overseas broadcast transmitters within their own territory (or, in the case of the BBC, in some cases in the countries of the Commonwealth).

The USIA planned its international broadcasting network around a number of strategic locations encircling the globe. America began to build a series of

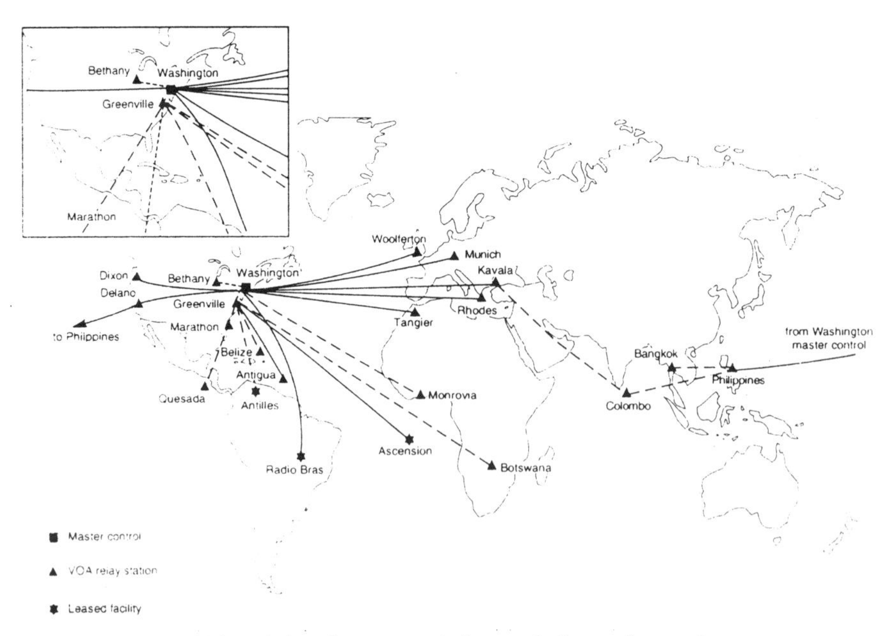

Figure 16.1 *Voice of America: transmission methods to relay stations*
——— *satellite,* — — — *shortwave,* - - - - - *microwave or land line,* ■ *master control,* ▲ *VOA relay station,* ★ *leased facility*

Figure 16.2 *VOA relay station at Tinang, Philippines*

airfields and military bases on the soil of its wartime allies. War-torn Europe, which had barely recovered from the devastation of world war, now found itself to be the site of bases for US warplanes. But this was not the only hardware being positioned. Using the same policy VOA secured its own bases from which to broadcast.

The first overseas extension of the VOA was the BBC transmitting station at Woofferton, England (although in this case the transmitters belonged to the BBC and were leased to VOA). Other transmitter bases were located in Tangiers, the Philippines, Okinawa, Kavala, Greece and Liberia within the first few years, while longer term plans were being made for even greater expansion.

The main advantage of this strategy was in the penetrative power it gave. A powerful radio transmitter located within a few hundred or even a thousand miles of its target zone can realise a stronger and more stable signal. The greater field strength can help to overcome any jamming that might take place.

The first major phase of the postwar expansion took shape in 1954. Poro in the Philippines was equipped with one MW transmitter and seven SW transmitters. The MW transmitter had a carrier power of 1000 kW, and combined with the SW transmitters made a total station output power of 1385 kW. Later the same year a station opened at Bangkok, Thailand, with a 1000 kW MW transmitter, and a new transmitting station on the Island of Rhodes, Greece came on air with 500 kW MW power and two 50 kW SW transmitters.

In the same year, the first phase of the Greenville project was completed. Greenville, in North Carolina, was the site of VOA's most ambitious project. Unlike other VOA stations in America, Greenville was planned and executed by VOA engineers rather than the private broadcasting companies. It had a dual function: it was to be the main VOA transmitting station on the US mainland, and it was to be the headquarters of the SW relay providing programme feeds to other transmitting stations.

There were three stations; Greenville A, Greenville B and Greenville C. Greenville A and B were identical in design and construction. Each had nine SW broadcast transmitters and two 10 kW single-sideband communications transmitters for programme relay and point-to-point communication. The SW broadcast transmitters were supplied by three different manufacturers; there were three 50 kW units manufactured by Gates Radio, three 250 kW units built by General Electric and three 500 kW units built by Continental Electronics. The Gates and GE transmitters used a design based on the classic class C amplifier chain, modulated at high level with a class B modulator.

The Continental 500 kW SW transmitters were of a completely different design. In what was certainly the highest powered SW transmitter ever designed, Continental had adopted a brand new method. The 500 kW transmitter was made up of two 250 kW high efficiency, linear amplifiers with Machlett type 5682 triode tubes, driven by a single 5681 tube modulated amplifier which was control-grid modulated by four RCA type 845 low-μ triodes in a parallel arrangement.

Yet another new feature of the Continental Electronics transmitter was the use of the Doherty method of high efficiency amplification in the 250 kW final stage. Equally advanced was the use of grounded grid operation, the tubes being in a driven cathode arrangement with the control grids grounded. This technique offered many advantages. The principle of the grounded grid goes back to the 1930s used in receiver circuit design, but its widespread application to high-power transmitter design did not come about until the late 1950s, when it was acknowledged as the state of the art in the renaissance period of the SW transmitter.

Greenville C was a SW receiving station located west of the Greenville transmitting stations, with enough distance between to allow satisfactory operation of the SW receivers in the presence of the white noise from the very high-power 500 kW transmitters. Greenville C performed a vital role in point-to-point communications and programme feeds to and from other transmitting stations, until the advent of satellite communications changed all preconceived ideas on programme relay methods.

The second phase of the postwar expansion of VOA's global broadcasting network came about in the mid 1960s. Bethany, Ohio, received three more 250 kW transmitters, Dixon transmitting station in California received three 250 kW transmitters, and Marathon, Florida was fitted with a 50 kW transmitter to serve Cuba. But the biggest expansion by far took place in Europe and Asia. In England, the Woofferton transmitting station was refitted with six new SW transmitters.

Quesada, Costa Rica, was fitted with a 50 kW SW transmitter to serve that country with the Voice of America. Colombo, Sri Lanka, was fitted with three transmitters to serve India. Africa was amply catered for by a giant SW transmitter complex in Monrovia, Liberia, which had six 250 kW units, and two 50 kW units. A second transmitting station in Africa came on air some several years later at Selebi-Phikwe, Botswana.

By far the largest VOA project planned for its global broadcasting network was for southeast Asia. As an addition to its existing high-power transmitting station at Poro in the Philippines, VOA built another giant transmitter complex in Tinang, Philippines. Tinang was equipped with three 50 kW SW transmitters in

1966, to be followed three years later with ten 250 kW SW transmitters. These were built by the Hughes Corporation. In more recent years another two 250 kW SW transmitters of Brown Boveri manufacture have been installed, making a total of twelve 250 kW transmitters. Also in southeast Asia, VOA installed a 1000 kW MW transmitter at Okinawa in the late 1960s, which has since been abandoned.

Tinang and Poro in the Philippines are central to VOA's southeast Asian operations. Tinang covers the regions of eastern Russia, China, southeast Asia and part of South Asia. Poro covers Vietnam, a densely populated region, with a 1000 kW MW unit, southwest China, southeast Asia and East Africa with its seven SW transmitters.

During the war years, and for at least another two decades, an underlying characteristic of VOA's operations was its reliance on the private broadcasting companies of the USA. Another feature of VOA was its policy of buying transmitters from every major manufacturer in the USA. Although this spirit of free enterprise and competition encouraged quality and was responsible for much innovation in design, it had its costs. For example, the sheer variety of transmitters increased training and familiarisation costs whenever staff were posted from one station to another. Some stations had as many as five different manufacturers' transmitters with their associated drive units, all different in design. These were the kind of problems that VOA Engineering and Technical Operations Division had to address for the future.

In the mid 1960s, the introduction of satellite communication links began to revolutionise the methods of getting programme feeds to SW stations overseas. The 1970s saw rocketing oil and electricity prices. This and the trend towards the use of more powerful transmitters signalled the need for a new generation of high efficiency transmitters.

Table 16.1 *VOA broadcasting sites as at February 1986*

Relay station	Primary programme feeds	Antennas	Transmitters	Power (kW)	Total power (kW)	Age (years)	Target area
Bethany, OH	Leased lines from Washington	22	6 SW	1×175	925	42	Latin America, West Africa
				3×250		17	
				2×175	350	42	(Used by AFRTS)
Delano, CA	Commercial satellite from Washington	15	9 SW	3×250	750	17	Central America, East Asia
				4×250	1000	1	
				2×100	200	41	(Used by AFRTS)
Dixon, CA	Commercial satellite from Washington	14	3 SW	3×250	750	17	Central America
Greenville, NC	VOA microwave from Washington	66	16 SW	6×500	4500	32	Latin America,
				6×250		22	North & West Africa,
				4×500	2000		Western Europe
Marathon, FL	Commercial satellite from Washington	1	1 MW	1×50	50	23	Cuba
Total US:		118	34 SW	34 SW	10475 SW		
			1 MW	1 MW	50 MW		
Judge Bay, Antigua	Commercial satellite from Washington	1	1 MW	1×50	50	17	Lesser Antilles to St. Lucia

Table 16.1 *Continued*

Bangkok, Thailand	SW from Philippines	1	1 MW	1 × 1000	1000	32	Southeast Asia
Selebi-Phikwe, Botswana	SW from Greenville and Bethany	1	1 MW	1 × 50	50	4	Northern S. Africa & Southern Zimbabwe
Kavala, Greece	Commercial satellite from Washington	23	1 MW 10 SW	1 × 500 9 × 250	2750	32 14	MW—Eastern Europe (Romania, Yugoslavia) SW—Middle East, S. Asia, South Central & Western USSR and Eastern Europe
				1 × 250	250	14	(Used by Government of Greece)
Rhodes, Greece	Commercial satellite from Washington	8	1 MW 2 SW	1 × 500 2 × 50	600	32 22	Arabic Middle East
Monrovia, Liberia	Commercial satellite from Washington	27	8 SW	6 × 250 2 × 50	1600	21	Sub-Saharan Africa
Munich, Germany	Commercial satellite from Washington	17	1 MW 4 SW	1 × 300 4 × 60	540	39 49	MW—Eastern Europe (Czechoslovakia, Poland, Hungary); SW—Western USSR and Eastern Europe
Poro, Philippines	Commercial satellite from Washington	17	1 MW 7 SW	1 × 1000 1 × 35 2 × 100 3 × 50	1385	32 32 32 22	MW—Vietnam SW—China, Southeast Asia & E. Africa
				1 × 35	35	32	(Used by AFRTS)

Table 16.1 *Continued*

Tinang, Philippines	Commercial satellite from Washington	39	15 SW	3×50	3150	20	Eastern USSR, China,
				10×250		17	Southeast Asia, and
				2×250		4	Eastern South Asia
Quesada, Costa Rica	Shortwave from Greenville	1	1 MW	1×50	50	17	Northern Costa Rica
Colombo, Sri Lanka	Shortwave from Kavala and Philippines	19	4 SW	2×35	80	32	India
				1×10		32	
				1×35	35	32	(Used by Government of Sri Lanka)
Tangier, Morocco	Commercial satellite from Washington	32	10 SW	3×100	440	35	Eastern Europe and
				4×35		40	North Africa
				1×100	200	35	(Shared with Government
				2×50		45	of Morocco)
Punta Gorda, Belize	Shortwave from Greenville	2	2 MW	2×50	100	1	Northern regions of Central America
Woofferton, England	Commercial satellite from Washington	37	10 SW	6×250	2700	22	Western USSR, Caucasus
				4×300		5	and Eastern Europe
Total—overseas			10 MW	10 MW	3550 kW		
			70 SW	64 SW	10 945 kW		
		225	80	74	14 495 kW		
Total—US & overseas			11 MW	11 MW	3600 kW		
			100 SW	90 SW	18 870 kW		
		343	111	101	22 470 kW		

Office of Engineering & Technical Operations (VOA/E), Voice of America, Washington, DC 20547

Chapter 17

Satellite communications and global broadcasting

'An artificial satellite at the correct distance from the earth would make one revolution every 24 hours; i.e., it would remain stationary above the same spot and would be within optical range of nearly half the Earth's surface. Three repeater stations, 120 degrees apart in the correct orbit, could give television and microwave coverage to the entire planet.'

Arthur C. Clarke wrote that in 1945, when rocket technology was in its infancy and only visionaries even thought of satellites. But the most remarkable thing about this statement is that it was made before the invention of the transistor by William Shockley and co-workers at Bell Labs. From its first demonstration in December 1947, the way was clear for a number of evolutionary and revolutionary developments in communications and broadcasting. A whole new science was born: solid-state electronics.

Electronics made great inroads into areas of technology that had stood still, relatively speaking, for many years. One of these was in the area of transoceanic and transcontinental telephone and telegraph communications. Another was the adaptation of solid state to those equipments that had previously depended on thermionic tubes. The first electronics market to be dominated by the transistor was the domestic consumer industry and by the mid 1950s the all-transistor, battery-operated radio receiver was being produced in millions.

The first artificial satellite was launched by the Soviet Union in 1957. The subsequent rate of development in satellite technology has confirmed that the launching of that first communications satellite was even more momentous than anyone supposed at that time. This first satellite sent radio signals back to Earth on a frequency of 31.5 MHz. In 1958 America launched its first satellite, and two years later another satellite was launched equipped with a passive reflector; it was used to relay telephony signals. These first two satellites, *Score* and *Echo*, left a lot to be desired but they were a necessary evolutionary step to the design of the active satellite, fitted with a receiver, a transmitter and transmit-receive antennas.

The first generation of active satellites travelled in a low-Earth orbit, and it was therefore a necessary part of the system to have Earth stations fitted with high-gain antennas that tracked the satellite as it passed over the Earth, and once the satellite had passed over the horizon all communication was lost until it

Figure 17.1 *4.5 m C & Ku satellite dish in a downlink installation*

reappeared. The solution to this problem was to use more than the one satellite.

The evolution of the geosynchronous satellite solved all these problems. If a satellite could be placed high above the Earth and travelling at such a speed that it appeared to be stationary with respect to Earth, this would provide continuous communications coverage to a large part of the Earth's surface. The first geosynchronous satellite was successfully launched on 2 April 1965 and placed into a geosynchronous orbit at 35 600 km above the coast of Brazil, proving the concept envisaged some 20 years earlier by Arthur C. Clarke.

This first geosynchronous satellite was limited in channel capacity to a mere 240 voice channels, but even so it was a dramatic improvement over existing methods of trans-Atlantic links. This satellite, *Early Bird* (more properly known as Intelsat I) revolutionised the communications industry in general, and radio broadcasting in particular. For the first time, live television pictures could be relayed across the Atlantic Ocean and shown on European television channels.

Although not highlighted by the media to anything like the publicity given to this feat, the satellite had an equal impact on radio broadcasting, and more particularly international broadcasting carried out on the high frequency spectrum, and it is only by looking at things in retrospect that the full impact of satellite communications can be properly appreciated. The first transatlantic relays, in the late 1920s, were primitive and often unreliable. The technique consisted of setting up a SW circuit from (say) New York to London over which a domestic programme going out over the US domestic networks would be relayed and simultaneously rebroadcast by the BBC to its listeners. Although the programme relay was unidirectional, it was also necessary to have a two-way communications link over which the engineers at each end of the link could converse in order to effect the line up.

One of the main requirements for a programme relay was a satisfactory SW link, but with the state of the art in the 1930s this was often difficult to achieve. The vagaries and problems associated with the ionosphere caused frequency distortion of the received signal and other difficulties such as fading, and at times a complete fadeout of the signal. Fading came in a number of flavours: slow fading, fast fading, flutter fading and sometimes a complete loss of the signal. To a certain extent the difficulties associated with achieving a good signal could be tolerated in commercial, military and marine communication systems, but not so with radio broadcasting. To effect an increase in radio circuit reliability from 60% to 80% called for a hundred-fold increase in transmitted power.

In an effort to increase the reliability of SW relays, higher-powered SW transmitters were brought into service by the mid 1930s together with high-gain SW directive antenna systems, which achieved circuit gains varying from 8 to 21 dB depending on type and complexity of design. Even so, it was quite usual in the late 1930s listening on the shortwaves, to hear frustrated line-up engineers in New York calling repeatedly 'Come in London, come in London' in efforts to establish a SW relay circuit. Often London could hear New York satisfactorily, but the communications link the other way had failed to be established, and without this engineering link the broadcast could not be properly lined at both ends as regards levels. Under such circumstances the relay would have to be aborted.

During the 1930s, throughout the war and even up to the 1950s there was only the one method of transatlantic telephony: the SW circuit. The poor reliability of this system posed many problems during the war—it was often easier for Churchill and Roosevelt to meet personally than to attempt long transatlantic discussions by radio telephone.

It was the operational requirements of the war that brought about one of the most costly and extensive projects ever undertaken in SW communications: the MUSA project. MUSA (multiple unit steerable array) was a desperate attempt to transform what was a fundamentally unreliable communications system into a reliable one. Shortwave broadcasting was then unreliable. There was nothing that could not be corrected given 30 more years of evolutionary development, plus the advent of computers, super transmitters and a better scientific understanding of the ionosphere, but in the meantime there was a war to be won.

MUSA aimed to combat the apparently random behaviour of radio signals that used the ionospheric mode of propagation. It was known that these signals were a combination of reflections by different modes; MUSA tried to combine the outputs from many different receiving antennas to produce a more constant signal. At Cooling Marshes, Kent, the British Post Office built a receiving system using 16 horizontal rhombic antennas. Each of the large rhombics was connected to low-loss coaxial cable, which transferred the signal output to the receiver system.

The 16 different outputs were adjusted in phase, and so combined as to yield

Figure 17.2 *Lift off. A communications satellite (Astra 1B) being launched into space orbit*

a signal that was relatively constant in amplitude, and with a substantial improvement in signal-to-noise ratio. A similar installation was built in the USA, at Manahawkin, New Jersey, operated by AT&T.

MUSA came into operational use in the first year of the Second World War. It played a vital role by making it possible for Roosevelt, Churchill and their respective chiefs of staff to hold vital discussions on the conduct of war while 5000 km apart. MUSA stayed in service throughout the war and for another 14 years, by which time the first transatlantic submarine telephone cable came into service. A few years later, the communications satellite, still in the developing

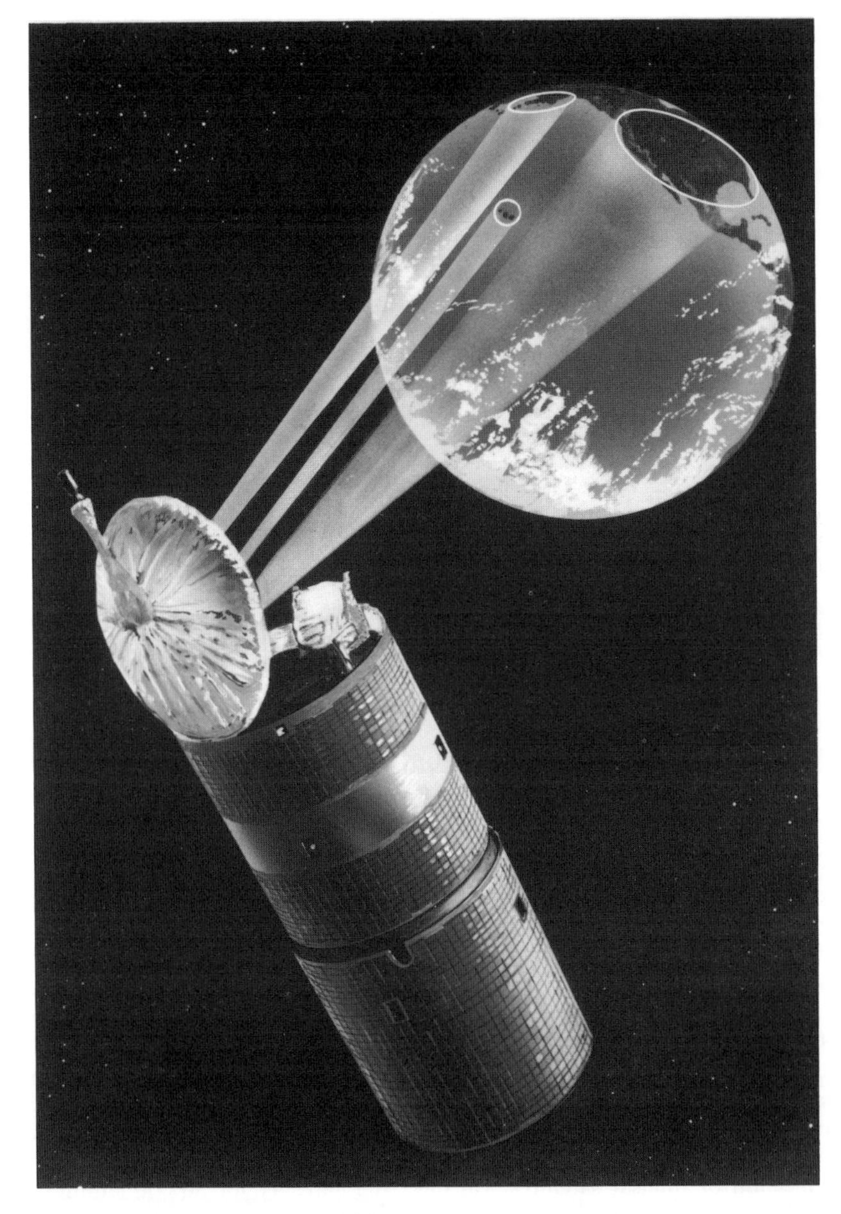

Figure 17.3 *Hughes' 'Galaxy' satellite*

stages, had proved itself to be the ultimate in reliable, trans-global communications. Suddenly MUSA was a thing of the past, and faded into history.

17.1 Satellites and broadcasting

Following the launching of *Early Bird* in 1965, making transatlantic communication

Figure 17.4 *Hughes' wide-bodied HS 601 satellite*

via satellite possible, a second satellite, *Intelsat II*, was successfully launched in 1967. This satellite connected the Pacific with America. Two years later Intelsat III was launched, covering the Indian Ocean. With all three satellites in service, Arthur C. Clarke's vision of more than two decades earlier had come true, and the way was open for a relay system with global capability to handle telephony, data, sound broadcasting and even video signals.

Establishing an international satellite communications system was only the beginning, however. Ongoing research and development into many aspects of satellite communications, providing greatly expanded services at lower operating costs with better reliability, continued (and still continues today). The sixth generation of Intelsat satellites will have the capacity for 30 000 simultaneous telephony circuits and three television channels.

But what has the satellite brought to international broadcasting? The use of the satellite link as a programme feed has made the dream of every broadcasting agency a reality. It has permitted VOA, the BBC World Service, Deutsche Welle and a half a dozen more international broadcasting agencies to place their radio transmitters many thousands of miles away from the main control room and studios, closer to the target country of propaganda broadcasts.

Although the programme may originate in Washington or London, the satellite link enables the programme to be broadcast from a station several thousand miles away. The BBC has been able to install two powerful shortwave transmitters on the shores of Deep Bay, Hong Kong, actually overlooking communist China, less than 5 km away. Senders 801 and 802 (as the giant 300 kW carrier power transmitters are known in BBC engineering circles) are now radiating a giant footprint over China, from Chunking in the west to the island of Hokkaido in the east, from Wuhan and Shanghai in the south, to Ulan Bator, Irkutst, and the valleys of Ob to the north. Most of the time this prolific station is deserted; the programme feed comes from Bush House, London, over a satellite link.

Part 3: Radio as an instrument of foreign policy

Chapter 18

Developments since 1960

The period from the early 1960s to the present day has seen a staggering growth and proliferation in international broadcasting, outpacing even the growth in the younger fields of VHF frequency-modulated sound broadcasting and television. International broadcasting is carried out by amplitude-modulated wave, the oldest and most basic form of radio broadcasting, using the high frequency band from 3 to 30 MHz.

Although it was undoubtedly the Cold War that fuelled interest by other nations in information broadcasting, not all information broadcasting is aggressive. The most common use today lies in its importance as an instrument of foreign policy. Cultural programmes can be used as a cloak for political propaganda of a much more subtle type.

The unique quality of SW transmission and reception is the ease with which information can flow, spanning oceans, bridging continents, crossing national frontiers, and reaching out to millions of listeners in any designated region of the world. For example, the BBC SW transmitting station in Rampisham, Dorset, England was built and designed to cover the whole of Eastern Europe and much of the Soviet Union.

Twenty-five years ago the advent of the communications satellite seemed to spell the end of terrestrial broadcasting. In particular one form of communication that was threatened more than others was the high-frequency spectrum—the short waves—which was also threatened by developments in tropospheric-scatter broadcasting. And yet, within another decade the HF spectrum had acquired an urgency and importance that nobody had foreseen. This was the realisation that the SW spectrum might be the only form of communication likely to survive an all-out nuclear war. In the case of SW broadcasting, as distinct from SW communications, interest in this was further stimulated by the continuing Cold War.

Paradoxically, it was the subsequent developments in satellite communications that gave a greater emphasis to SW broadcasting. The communications satellite was such a powerful adjunct to SW broadcasting that the combination of the two created a synergetic force. SW broadcasting assumed its trans-global capability to a degree never before possible. For the first time, it became possible for a programme coming out of VOA's Washington studios to be simultaneously transmitted by satellite link to the SW transmitters around the world, relayed to millions of listeners.

Before the introduction of satellite communications programmes were relayed by a variety of methods, but over long distances a SW relay was used. As late as the early 1960s, when VOA was broadcasting on short waves to the Soviet Union from its HF relay station at Woofferton, England, the programmes had been sent from Washington over a SW link. This system of transmitting a programme feed from the central studios to a SW transmitter a few thousand miles distant could never guarantee a 24 hour reliability.

The communications satellite can give that guarantee, and it is this reliability that has enabled SW international broadcasting to develop as it has. Fifty years ago the BBC World Service had just one short wave station at Daventry, England; now it has a network of HF relay stations. Today it has transmitters in the Caribbean and Ascension Island, in the Indian Ocean it has Masirah Island, in the Far East it has relay stations in Krangi, Singapore and, most recently, Tsang Tsui, Hong Kong.

Tsang Tsui is the BBC's East Asia relay station, equipped with two 250 kW SW transmitters of Marconi manufacture. It is one of the few SW stations in the world designed to operate without human supervision. With its high-gain curtain arrays, this station will project a powerful signal to Japan, Korea and especially to the People's Republic of China. Quite why the UK Foreign & Commonwealth office should provide a budget of £100 million to build a radio station in a colony that the British have to give up to China by 1997 is anybody's guess. One could possibly conjecture that the intention is to bring about an internal uprising in China with the hope that the present regime will be overthrown, thereby averting the takeover of Hong Kong.

Information broadcasting is and will remain a political instrument of foreign policy. Because of this fact, countries have tended to align their foreign broadcasting into 'east and west' political camps. In the Far East, there is rivalry between the People's Republic of China and the Chinese Republic on the island of Taiwan. Taiwan relays propaganda programmes originating from the USA, through station WYFR in Florida. But, for all the expenditure on information broadcasting by America, Western Europe and the USSR the greatest growth rate of all has occurred in the Middle East by the Arabic-speaking nations.

The development of broadcasting in the Arab world has been led by Iraq, Iran, Jordan, Kuwait, Saudi Arabia and Libya—although there is not a single Arab nation that has not invested in the latest technology for super-power broadcasting. In terms of investment in super-power broadcasting per head of population, the highest expenditure was by Kuwait.

Within the past decade another emergent group in information broadcasting is in privately owned and operated religious broadcasters. Religious broadcasting began in 1927 from the Vatican, followed in the late 1940s by Trans World Radio. These two religious broadcasters are well known and respected. The growth in the past decade of private broadcasting agencies seems to have more to do with world politics than pure religion.

Religious broadcasting shows every sign of being the next generation of political propaganda broadcasting, taking over from the accepted style of international broadcasting. Already, no fewer than three religious broadcasters in the USA have acquired 500 kW SW transmitters, and one has recently put into service two

500 kW units in a phased arrangement, to produce 1000 kW of carrier power.

Although the rapid growth of SW broadcasting continues, the medium-wave band has now acquired a significance in international broadcasting. From being the traditional preserve of national and regional broadcasting, the MW spectrum has become favoured by international broadcasters. The growth in international MW broadcasting is now approaching the popularity of the short waves, but for different reasons.

SW broadcasting is still the only medium that can reach out to audiences several thousands of kilometres distant, with the assurance that reception cannot be censored or prevented. Some countries, outside Europe, use the short waves for domestic broadcasting. Generally, these are the countries with huge land masses, such as China and the USSR. Thus there are more SW receivers to be found in these countries than in any other region of the world; SW broadcasting will therefore continue to be the main vehicle for information broadcasting directed to these countries from the West.

In Europe, however, there has been a significant change in reception trends over the past two or three decades. From what was a popular pastime 50 years ago, SW broadcasts are now listened to by only about 25 out of every thousand listeners, while others turn their attention to television or VHF-FM. This trend has not gone unnoticed by the broadcasting agencies or the receiver manufacturers. Today very few new radio receivers are fitted for SW reception; but the MW band is to be found in practically every household in Europe, including Eastern Europe and the Baltic States.

The medium waves possess advantages and certain properties that make them eminently suitable for international broadcasting. The MW band extends from 540–1600 kHz. The upper end of this frequency band exhibits properties similar to those found on the SW bands: after dark the broadcast reception range is considerably extended because of ionospheric reflection. As a result, the MW band is rapidly becoming the spectrum for international broadcasting, while national broadcasting migrates to VHF-FM. There are good reasons for the increasing popularity of VHF for national broadcasting: transmitters are lower powered, antennas require less ground area and sound quality is better. However, no financial constraints apply to international broadcasting.

There are now 22 LW and MW transmitters with output powers of 2000 kW: eight of these are in Europe, the other 14 in the Arabic-speaking countries of North Africa and the Middle East. Broadcasting with megawatts on long and medium waves calls for a vast capital investment, which few countries can afford, and the running costs of such a station equate to the electricity consumed by a town.

For these reasons, super-power broadcasting in this league will be a growth area of broadcasting in the rich oil-producing Arab world where, because of a common language and culture, it is convenient to broadcast the same programme for national audiences and to adjacent countries. Programmes broadcast from Cairo can be heard all along the North African coast, as far west as Morocco, and eastwards in Saudi Arabia.

Voice of America also uses some 1000 kW MW transmitters for broadcasts aimed at high-density audiences in the Far East. VOA has two such transmitters: one in Bangkok, the other in the Philippines. Both are used for propaganda broadcasting to communist southeast Asia.

Super-power broadcasting by the Arabic countries and super-power broadcasting with megawatts on the medium waveband are the two main growth areas of terrestrial broadcasting for the dissemination of information-type broadcasting into the 21st century. Because of their importance both are covered in later chapters.

Yet another growth area in broadcasting for the next century will be increased investment in multi-frequency, multi-band, fully automated broadcasting complexes. Again, the chief investors will be the rich oil-producing Arab states.

Chapter 19

The decade of audibility: 1980–1990

The 1980s saw a proliferation in super-power broadcasting, as ever more countries sought stronger voice on the radio spectrum. In the 1970s broadcasters such as Radio France International and Deutsche Welle had acquired eleven and nine 500 kW transmitters, respectively.

In 1984 the Reagan administration declared that the USA would not compromise in its intention to maintain a free flow of information across territorial and international borders, using the SW bands. That same year the US Information Agency (USIA) issued tenders to the five major manufacturers for an initial number of 55 SW super transmitters of 500 kW carrier power.

Such has been the rate of growth in information broadcasting on the HF spectrum since the early 1980s that today over 25 countries have acquired, or ordered, SW super transmitters.

19.1 Broadcasters

Over 150 countries carry out broadcasting on the AM wavebands on a national basis. Of these, over 100 countries also carry out foreign broadcasting on the AM wavebands, mostly on the SW band but with a growing interest in the MW band. The most up-to-date information on terrestrial-based broadcasting, be it AM, VHF-FM or television broadcasting, is to be found in the *WRTV Yearbook*, published annually and updated to take account of the latest available information on countries and their acquisition of new transmitters.

However, even WRTV is not necessarily up-to-date at the time of its publication. It may not include details of planned acquisitions by certain countries, and there are those countries that will exhibit a degree of reticence concerning the issue of what it regards as sensitive data on its foreign policies.

Other sources of information concerning the world's broadcasters are the BBC and USIA. Both of these broadcasters issue data on the world's leading countries in information broadcasting on the AM wavebands. The data published by the International Broadcasting and Audience Research Unit is particularly useful. It lists the top 30 international broadcasters in terms of countries and their weekly output in programme-hours per week, from which may be deduced the importance attached by each country to foreign broadcasting. The chart is also comprehensive in its statistics back to 1950, which show the growth that has taken place over the past four decades.

According to this chart, the top five countries in propaganda broadcasting are the USA, the USSR, China, Germany and the United Kingdom. Growth in propaganda broadcasting has been three-dimensional: in numbers of countries, in numbers of broadcasters per country, and in numbers of transmitters. The USA is acknowledged to be the world's largest international broadcaster, with over 2500 programme-hours per week in 1986 (the sum total output from VOA, Radio Free Europe and Radio Liberty). The US lead is to be consolidated over the next decade as VOA brings onstream its new generation of multi-frequency, super-power HF stations as exemplified by sixteen 500 kW SW transmitters, now being planned, to be located in the Negeb desert in Israel.

In second place, marginally behind the USA, was the USSR, and in third place the People's Republic of China, with a weekly output of 1446 hours. In fourth place was West Germany with 821 hours, followed by the UK (BBC) with 733 hours. It may be assumed that the latest figures are higher by about 15%. Relative to its huge land mass, the weekly output quoted for China is low compared with that of the UK and Germany, which shows the high importance attached by these two countries to propaganda broadcasting as an instrument of foreign policy. Currently, both countries are expanding their international broadcasting capabilities. Deutsche Welle, the foreign broadcasting service of Germany, is expanding its investment in 500 kW SW transmitters, and will shortly become the most powerful broadcaster in the western world outside the USA.

The BBC World Service has a powerful advantage over any other country—the size of its listening audience. No other country has such a universal appeal to listeners around the world; not even America or the Soviet Union can match the attraction that the BBC offers to its listeners.

The BBC's popularity is based on a mixture of fact and myth. It is one of the oldest foreign broadcasters in the world, along with Germany, Italy and the USSR, with its foreign broadcasting dating from the late 1920s—although it broadcast to the world in the English language only until 1938. The myth is the idea, successfully fostered by the BBC for almost 70 years, that it is independent of government control or censorship. So successful has been the implantation of this idea that it is believed even by most of its staff. To sustain this myth the BBC uses a number of techniques, including the 'window of credibility': from time to time, the BBC broadcasts news that is calculated to embarrass the government. The technique is very successful, and has been initiated by other international broadcasters.

In making any sort of analysis of the chart of international broadcasting outputs of various countries prepared by the Audience Research Department of the BBC, one should bear in mind that it is a document that is intended for worldwide publication. It is therefore selective in its treatment, and there are some glaring omissions of statistics relating to broadcasting.

No mention is made of the fact that many countries have more than the one government-owned broadcasting agency. Sometimes it is politically desirable to have an overt voice and a covert voice on the airwaves. In the US the broadcasters shown separately from VOA are Radio Free Europe, Radio Liberty and

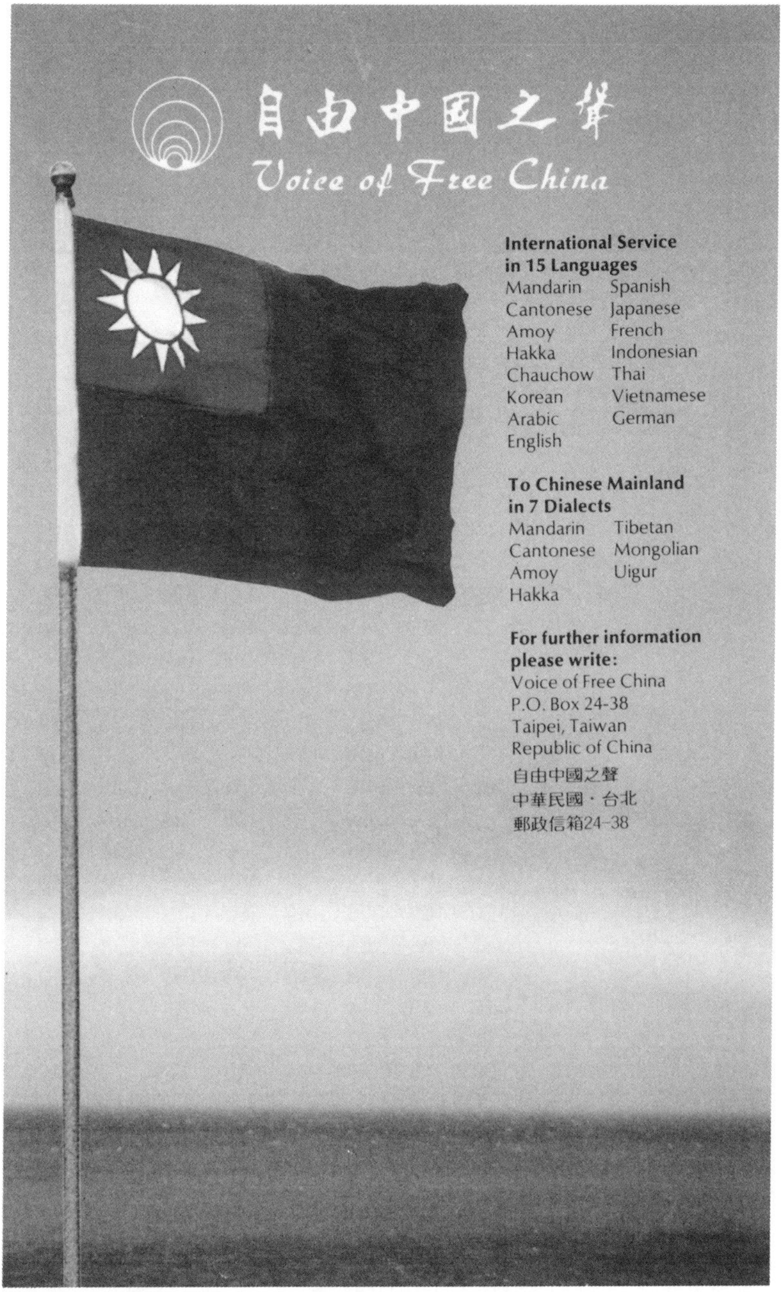

自由中國之聲
Voice of Free China

International Service in 15 Languages
Mandarin
Cantonese
Amoy
Hakka
Chauchow
Korean
Arabic
English
Spanish
Japanese
French
Indonesian
Thai
Vietnamese
German

To Chinese Mainland in 7 Dialects
Mandarin
Cantonese
Amoy
Hakka
Tibetan
Mongolian
Uigur

For further information please write:
Voice of Free China
P.O. Box 24-38
Taipei, Taiwan
Republic of China
自由中國之聲
中華民國・台北
郵政信箱24-38

Figure 19.1 *Advertisement in WRTH for Voice of Free China*

Table 19.1 External broadcasting *Estimated total programme hours per week of some external broadcasters*

	1950	1955	1960	1965	1970	1975	1980	1985	1986
United States of America	497	1690	1495	1832	1907	2029	1901	2339	2411
USSR	533	656	1015	1417	1908	2001	2094	2211	2229
Chinese People's Republic	66	159	687	1027	1267	1423	1350	1446	1446
German Federal Republic	—	105	315	671	779	767	804	795	821
United Kingdom (BBC)	643	558	589	667	723	719	719	729	733
Albania	26	47	63	154	487	490	560	581	588
Egypt	—	100	301	505	540	635	546	560	560
North Korea	—	53	159	392	330	455	597	535	535
East Germany	—	9	185	308	274	342	375	413	446
India	116	117	157	175	271	326	389	408	408
Cuba	—	—	—	325	320	311	424	379	381
Australia	181	226	257	299	350	379	333	352	359
Iran	12	10	24	118	155	154	175	310	324
Nigeria	—	—	—	63	62	61	170	322	322
Poland	131	359	232	280	334	340	337	320	320
Netherlands	127	120	178	235	335	400	289	336	316
Bulgaria	30	60	117	154	164	197	236	290	315
Turkey	40	100	77	91	88	172	199	307	307
Japan	—	91	203	249	259	259	259	287	282
France	198	191	326	183	200	108	125	272	274
Czechoslovakia	119	147	196	189	202	253	255	268	267
Spain	68	98	202	276	251	312	239	252	267
Israel	—	28	91	92	158	198	210	223	220
Romania	30	109	159	163	185	190	198	212	208
South Africa	—	127	63	84	150	141	183	205	205
Italy	170	185	205	160	165	170	169	173	173
Canada	85	83	80	81	98	159	134	169	171
Portugal	46	102	133	273	295	190	214	140	155
Sweden	28	128	114	142	140	154	155	196	154
Hungary	76	99	120	121	105	127	127	122	122
Yugoslavia	80	46	70	78	76	82	72	86	86

(i) USA includes Voice of America (1226 hours per week), Radio Free Europe (566), Radio Liberty (497) and Radio Marti (122) (1986 figures).
(ii) USSR includes Radio Moscow, Radio Station Peace & Progress and regional stations.
(iii) German Federal Republic includes Deutsche Welle (568 hours per week) and Deutschlandfunk (253) (1986 figures).
(iv) The list includes fewer than half the world's external broadcasters. Among those excluded are Taiwan, Vietnam, South Korea, and various international commercial and religious stations, as well as clandestine radio stations. Certain countries transmit part of their domestic output externally on short waves; these broadcasts are mainly also excluded.
Source: International Broadcasting & Audience Research.

Radio Marti. Yet it is now common knowledge that these three latter broadcasters are government funded. Similarly, no mention is made of the fact that until the mid 1980s Britain operated more than the one broadcasting authority. The British Foreign Office maintained its own high power shortwave and medium wave broadcasting stations in Britain and in other parts of the world.

Another notable exception, for which no reason is given, is the absence of any broadcasting statistics on Taiwan. This country, which calls itself the Republic of China, has a broadcasting capability which on a per capita basis is possibly the largest in the world. It has four broadcasting agencies: the Central Broadcasting System (CBS), the Broadcasting Corporation of China (BCC), the Voice of Free China (VOFC) and WYFR Family Radio, which is a relay from WYFR Florida. Much of the broadcasting output from these different agencies is beamed across the short span of the China Sea which separates Taiwan from Mainland China. There is little secrecy elsewhere about the fact that Taiwan, the Republic of China, is dedicated to the overthrow of the communist regìme on mainland China, and there is equally little doubt that it is being assisted in these endeavours by the West.

If the chart of international broadcasters were rearranged on a per capita basis, the results would be quite different. Leading the table would be some Arabic-speaking countries, and Taiwan and the UK would also be near the top. Because the measurement parameter is hours of programming, the chart is deceptive in some ways. For instance, Arab countries tend to radiate the same programme over different transmitters, making the output appear to be low, whereas this is far from being the case.

One of the most respected and influential broadcasters is France, yet it lies two thirds of the way down the list. France broadcasts in fewer languages than VOA, Radio Moscow and the BBC. France has two international broadcasting agencies: Radio France International, the main outlet for SW broadcasting to the ex-colonies, and Radio Monte Carlo, jointly owned by the governments of France and Monaco. Radio Monte Carlo exerts considerable influence in North Africa and the Near East, and functions as a culture-driven broadcasting agency rather than politically driven.

Chapter 20

Technology of the high-power transmitter

The modern AM transmitter has an evolutionary path back to the first transmitters of the early 1920s, with a few revolutionary developments along the way. There are four distinct phases of development: the quest for higher power (up to 1939), the first post-war phase (1948–1958), the second post-war phase (1959–1979) and the quest for higher conversion efficiency (1979 onwards).

In each phase the transmitting tube played a vital role. The tube is the most vital component in a transmitter, and up to the 1960s it was the limiting factor in the generation of very large amounts of RF power. Today, tube development has reached a stage where the voltage stresses and RF current densities produced in the final output stage of a modern high-power SW transmitter impose tremendous stresses on other components, notably the RF switches, the point contacts in the tuning mechanism of the final tank inductors and, perhaps most important of all, the high-vacuum variable capacitors.

20.1 Chronology

Up to 1939, radio broadcasting was an immature technology and there was much that needed to be achieved. Radio broadcasting in the USA was a competitive business, where the output power from a transmitter was deduced as the limiting factor to the range up to which the broadcasts from a station could be heard. Because radio broadcasting in America became driven by advertising revenues, and because the sponsors favoured radio stations with the loudest signal, it was inevitable that there would be a race for power output. Thus power outputs went up from watts to hundreds of watts, and finally tens of kilowatts. With a few notable exceptions—such as station WLW, which got a licence to build a 500 kW MW transmitter in 1934—transmitter powers were between 20 and 50 kW by the late 1930s, with each radio station serving a region or state.

In Europe, practically every country had placed radio broadcasting under the control of the state, with a few exceptions such as France where radio stations were permitted to be used for advertising goods and services. Because European countries were much smaller than America, it was convenient from the start to think in terms of transmitters with sufficient power to cover provinces, and later the entire country. By the late 1930s Europe had many more high-powered broadcasting stations than North America.

Europe also had many more transmitter manufacturers. These companies were multi-disciplined, transmitters being just one of many different products. Other products included the tubes themselves. By the late 1930s Europe had several large companies involved in such sciences, all able to trace their origins back to the early days of electricity and electrical products—such as the incandescent filament lamp, the forerunner to the diode and the triode. These companies included Marconi and STC in the UK, Brown Boveri in Switzerland, Telefunken and Siemens in Germany, Thomson in France, Tesla in Czechoslovakia, Philips in Holland. Although radio broadcasting was controlled by states, it was national pride that acted as the spur to technology for national broadcasting, and by 1939 there were several broadcasting stations in Europe with output powers as high as 150 kW, one of which was Beromunster, Switzerland. (Only Radio Luxembourg aspired to 200 kW.)

All national and regional broadcasting in Europe then used the medium and long waves. The countries that became interested in short-wave international broadcasting were those with colonial outposts. SW transmitters thus served the colonial interests of nations such as France, the Netherlands, Italy, and political interests on the part of Germany and the Soviet Union, both acting as a spur to the development of SW broadcasting capabilities. To meet the needs of the global distances to be covered to reach Britain's colonial outposts, Marconi and STC began the development of high-powered broadcast transmitters. Marconi produced the 75/100 kW SW transmitter type SWB-18, and STC produced its CS-8, a 50 kW SW transmitter.

Marconi's SWB-18, and the models of the same series that preceded it, represented everything good about British engineering. They gave the appearance of having been built to last a hundred years, and only a few years ago it was known that some were still in service in a few developing countries. These transmitters were constructed mainly from heavy gauge brass and copper sheet.

The SWB-18 was developed primarily for the BBC external SW services and went into service during the early war years as part of the British plans to increase its overseas broadcasting capability. It had a frequency range that extended from 3.75 MHz to 22 MHz. Up to 10 MHz the maximum permissible output was 100 kW; above 10 MHz the output was reduced to 75 kW. This restriction on power output at the higher frequencies was a characteristic of SW transmitter design up to the 1960s, due to excessive RF losses at the high frequencies, and the valve capacitances caused by the large copper mass of the anode.

Channel changing—the time taken to shut down the transmitter and retune it to another frequency—rated high in design priorities, but whereas the modern SW transmitter can do this in a few seconds, the same task 50 years ago could take two shift engineers the better part of an hour. The SWB-18 design team came up with the ingenious idea of tank inductors that could be wheeled into position to replace the one in use. Frequency changing was achieved by shutting down the transmitter, wheeling out the tank inductor and replacing it with another for the new frequency. In effect, the transmitter had its own railway lines to the coil storage area, with its own marshalling yard.

At the same time there appeared the RCA 50 kW transmitter, of which the BBC External Services purchased six. Also current was the STC CS-8, a dual-channel 70 kW carrier power transmitter. Since the parent company of STC was American owned, the CS-8 represented transatlantic design, and was quite different to the more sturdy SWB-18 in appearance. The BBC's overseas services during the war and for many years after was based on these three transmitters.

The first major British development in post-war broadcast transmitter design came with the Marconi BD-253. This was a 100 kW carrier power model designed for fast channel changing, achieved by the use of dual RF stages with a single modulator. Although it still took the better part of an hour to align the transmitter on to the two frequencies, thereafter all that was necessary to change to the other frequency was to switch RF drive and modulation to the second channel.

BD-253's were installed at BBC Woofferton and other transmitter stations, and were a mainstay during the Cold War. They were also sold to foreign broadcasting authorities. Shortly after the BD-253 entered service, the Marconi Company began development of a 250 kW transmitter: the BD-272. It was similar in design to the BD-253 but produced a higher carrier power. The BD-272 was another success story for Marconi: it sold some 30 units, compared with 23 models of the BD-253 that were sold. The BD-272 first came into service in 1964, and was sold to the BBC and many foreign-broadcasting authorities, including Deutsche Welle.

The overall efficiency of the BD-272 was similar to the BD-253, which, at 47%, was a good performance for that era. While the BD-272 was in service, Marconi

Figure 20.1 *Marconi BD-272 transmitters installed at the Woofferton VOA relay station in 1963*

was working on the BD-6122, a 300 kW design.

The BD-6122 was similar in design to the BD-272, but its overall efficiency was better and physically it was smaller. By this time (the late 1960s) transmitters were becoming more powerful but occupied less space. The BD-6122 was still a manually tuned transmitter, so that frequency changing was a time-consuming operation, still taking two shift engineers for most of one hour to change inductances in the output stage and realign the driver stages.

In the late 1970s Marconi brought out the B-6124. This was a 300 kW carrier power model, but it was a radical departure in design philosophy from anything that had gone before, in that it embodied servo-controlled, motor-driven variable capacitors to set the tuning of 36 different channel frequencies. With these channel settings stored in a digital memory, all that was necessary to change channel frequency was to press a button.

Tuning time was now reduced to the order of a minute. The B-6124 differed from the BD-6122 in one other important aspect: it took advantage of the very latest in tube technology, the Thomson-CSF Pyrolitic grid construction, which had made it possible to produce tubes that could handle hundreds of kilowatts of RF power. The first B-6124 was built in 1977, and the last was sold in 1981.

In the late 1960s, Marconi lost its lead in sales of high-power SW transmitters, to other European companies, Brown Boveri and Thomson-CSF. This was due

Figure 20.2 *Pyrolitic construction, machined grid for a tetrode transmitting tube*

to the advances in transmitter technology being made by these companies.

In the early 1970s, two major happenings had a profound effect on the major Western international broadcasting agencies and the manufacturers of transmitting hardware. These were the Cold War and the oil crisis. The continuing Cold War had entered a dangerous phase of escalating antagonism, as the Western nations increased the power of their broadcasts, and the Soviets retaliated by jamming operations. With no end in sight, the USA, UK and West Germany planned for more powerful transmitters than the 250/300 kW transmitters already in service.

In 1972 Thomson-CSF produced the world's first 500 kW SW transmitter. But higher powered transmitters make heavier demands upon electrical energy, requiring as much as 1000 kW from the electricity supply. This period coincided with the oil crisis, which had a traumatic effect on Western economies: as industries went onto part time, some broadcasters shut down some non-essential stations due to a crisis of supply.

The oil crisis passed, but the higher price of fossil fuels did not. The new energy costs focused attention to the need to conserve energy. In the broadcasting world the advent of the better and more powerful transmitting tubes enabled 500 kW to be produced from a single tube. These remarkable developments in power grid technology were the result of massive and sustained research and development by Thomson-CSF, and it was the broadcast division of that company that produced the world's first 500 kW SW transmitter.

The electrical conversion efficiency of this first generation of super transmitters represented a considerable improvement on anything that had gone before: overall efficiency had been lifted to nearly 60%. Even so, this still meant high operating costs, particularly where 500 kW transmitters were concerned.

20.2 Pulse-duration modulation

Since the early days of broadcasting, successive generations of engineers have applied their efforts towards increasing the basic efficiency of transmitters. Most of this work had been directed to the modulator section and to replacing the classic class B modulator stage with a more efficient method. The principles of pulse duration modulation were known, but had hitherto not been applied to very high power. The US company Harris was the first to produce a 100 kW transmitter embodying PDM techniques, and a number of these were sold to broadcasters such as VOA; the British Foreign & Commonwealth Office (FCO) is also believed to have bought some.

In 1982 the German company AEG produced the world's first 500 kW PDM transmitter. This brought about a transmitter with an overall efficiency of 65%. Other manufacturers quickly followed with similar models; Marconi and Thomson-CSF. In 1984, the Swiss company Brown Boveri, which had been pursuing its own line of research based on the use of semiconductors and fast-acting, solid-state switching devices, made a giant leap forward in transmitter design, with its pulse-step modulator (PSM). In place of the tube used by other manufacturers as the device for pulse-duration modulation, Brown Boveri engineers used a system of 32 power supplies. These individual outputs were connected in series via fast-acting switches.

At the peak condition of 100% modulation, all 32 supplies connected in series produced a voltage of 28 kV; at the modulation trough only one of the outputs would be used. This voltage applied to the RF tube to be modulated corresponded to the conditions for amplitude modulation where the plate voltage swings between $2E$, and zero during modulation, where E is the normal voltage applied to the tube during carrier-only condition.

The beauty of PSM is that it eliminates the lossy components such as transformers, chokes and filament-heater supplies. Because of this, the PSM technique realises an overall efficiency of better than 70%, which produces a very significant power saving in a transmitter hall with several 500 kW transmitters.

The pressing need for energy conservation, coupled with the need to reduce operating costs of powerful transmitter complexes, stimulated the search for more ways and means of increasing efficiencies. AEG and Brown Boveri produced a method for saving energy in the carrier, which contains two-thirds of the power in a normal AM signal although its only purpose is to assist in the demodulation of the received signal. Dynamic amplitude modulation (DAM) was the system developed by AEG, while Brown Boveri called its system dynamic carrier control (DCC). In essence they amounted to the same thing: varying the power in the carrier, in proportion to the depth of modulation.

Since the birth of broadcasting the classical mode of transmission has been amplitude modulation, and the indications are that this method will stay with us well into the next century. However, two thirds of the power is in the carrier, and can for all practical purposes be regarded as wasted except for the fact that the carrier is an essential component in the modulation and demodulation processes. At very low music passages and during speech pauses there is very little energy in the sidebands, so an obvious saving can be made by varying the level of carrier more or less in sympathy with the modulation. The drawback is that the received signal will be weaker during these periods.

The subject of varying the carrier power to save energy is not new—the idea dates back to the early 1930s, when it failed to meet with much enthusiasm for two reasons: the lack of the technical means to make the idea work in practice, and the fact that in those days no-one thought it necessary to save energy. In the 1930s the idea was called the Hapug principle, after the names of its three inventors.

More recently, however, with modern energy costs and the trend towards much higher powered transmitters, the idea of devising energy saving has gained prominence, and nearly all the major manufacturers have devised their own particular methods. Many transmitters are now in service with energy-saving devices fitted. The BBC uses a circuit which operates in the opposite way to any other broadcaster: the carrier is varied such that during speech pauses the carrier is at its highest level, thus preserving a good signal at the receiver. As the programme level increases the carrier power is reduced, but the worsened signal-to-noise level is masked by the greater volume output from the increased depth of modulation.

The method pioneered by ABB and AEG has received more universal acceptance. In this method the carrier varies in proportion to modulation depth, down to a predetermined minimum level (– 6 dB has been accepted as the best compromise).

The fitting of energy saving devices such as these has now become the number one priority for most international broadcasters; as output power from a transmitter rises, so does the degree of the importance of some form of DCC. At the time of writing, ABB is carrying out modifications to the high-power transmitter of Saarlaendische Rundfunk.

20.3 Speech processors

Recent years have seen some notable developments in speech processing devices. The principle of speech clipping in radio transmissions goes back several decades. Clipping—the process of removing peaks of modulation— has the effect of increasing the overall average level of the transmitted signal, at the expense of some very undesirable effects such as distortion, sideband splatter and harmonics. Because of these limitations, speech clipping found its main applications in communications rather than broadcasting.

The modern speech processor is sophisticated in design and performance; it can improve speech perception, help in punching through static, and partially counter the effects of jamming. As well as these advantages, the speech processor can save on energy by eliminating excessive bass response, which can otherwise consume power during such passsages (eliminating excessive bass response also extends the life of a transmitting tube).

Thus, in one way or another, the speech processor makes important contributions to the performance of modern high-power transmitters and for this reason it has assumed a growing importance in SW broadcasting. Orban, a San Francisco-based company, which manufactures the audio processor type 9105A, designed specifically for SW broadcasting, can claim the distinction of having supplied broadcasters in both East and West camps: Radio Moscow, VOA, the BBC and many others.

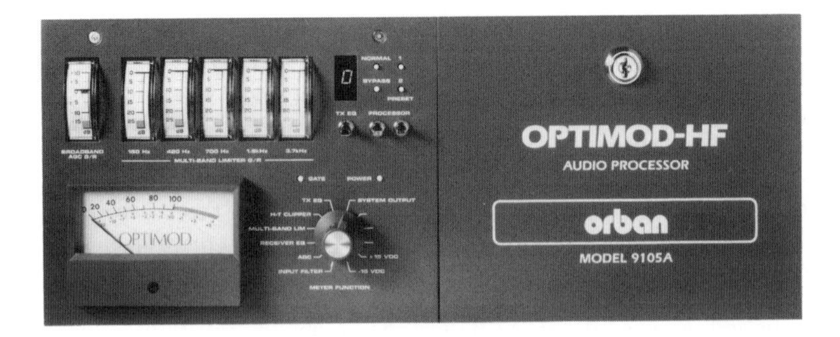

Figure 20.3 *Optimod-HF speech processor used by most of the world's broadcasters*

20.4 Rotatable curtain antennas

The rotatable curtain array is a versatile tool, which has come into prominence in the past decade. Traditionally, antennas for SW broadcasting have been of the fixed type; the one most commonly used was the fixed curtain array, an arrangement of dipoles stacked in a vertical and horizontal configuration, which, depending on the number of stacks, can realise a gain in the directivity from 17 to 23 dB.

In the late 1960s, the advent of the logarithmic-periodic antenna and its application to HF communications was later followed by its application to SW transmitters with powers of up to 100 kW or so. The horizontally polarised log-periodic, with its single-mast support, was found to be suitable for a rotatable arrangement, either by motor-driven or by a manual arrangement of steering it to the new angle of fire. A few companies succeeded in building log-periodic antennas of the motor-driven rotatable type for SW broadcasting, but the design did not really take off because of problems of mechanical stress and an electrical nature caused by very high voltages in the feeder and elements.

By this time the curtain array had become universally popular with SW broadcasting authorities, so it was logical that a rotatable version would follow. Practically all SW broadcasting relies on the use of directive antennas on a fixed azimuth bearing; to cover the world, therefore, a significant quantity is required. For example, a station with 15 SW transmitters might have as many as 36 directive antennas, and often many more. This immediately raises the problem of land required. Most of the BBC World Service stations, for example, have antenna sites that extend over several square kilometres; and in Western Europe generally, land is an expensive resource.

One antenna can only provide continuous transmission service from one transmitter, therefore for the broadcasting authority with only one or two transmitters the rotatable antenna is a cost-effective solution, able to cover different zones of the world at different times. Equally, the rotatable has advantages for large broadcasters who might have 20 SW transmitters, in that one rotatable can act as a backup for any one of the curtains on fixed bearings.

Disadvantages of the rotatable curtain array are its complexity of mechanics and the cost, which is very high compared with any fixed antenna. With increasing power handling these problems become magnified. As a result, there are few companies in the world with the necessary experience; those that have met with the greatest success in this field are ABB, AEG and Thomson-CSF. AEG has been successful in selling its version to European broadcasting authorities; one in the Netherlands, at Kvitsoy NTA, is probably the most northerly such station in the world. Others have been supplied to the Vatican City and ORF Austria.

ABB and Thomson-CSF have been similarly successful in supplying their versions to various countries in the Middle East, including Iraq, Iran, Kuwait, Turkey and a number of Gulf states. The rotatable curtain array is another product where the different companies have adopted quite different solutions to meet the same operational requirement. The AEG design is a structure of two towers, which can rotate through the full 360° by means of a circular railway track.

The ABB design is a Y-shaped structure from which the curtain array is suspended. A single column, which rotates on a roller bearing, carries the complete Y structure. This design gives joltless start and stop, and continuous acceleration and deceleration, with a positioning accuracy within 1°. The Thomson-CSF design is similar to the AEG and ABB versions in certain respects; its unique feature is that one side of the (otherwise perfectly balanced) structure is weighted. Thus,

Figure 20.4 *Rotatable curtain array for 500 kW shortwave broadcasting. Vatican City, Radio Vaticana, Santa Maria di Galeria*

in the event of a gale-force wind, the structure rotates so that its main axis is parallel with the wind direction. The clutch is designed to release when windspeeds exceed a certain figure, at which time the transmitter is shut down. The massive size of these rotatable curtains—of the order of 100 m by 100 m high—means a large surface area exposed to wind forces.

Possibly because of the high costs of rotatable curtains there are at present no more than 20 or so countries equipped with them, but they certainly represent a current growth area in SW broadcasting.

20.5 Single-sideband broadcasting

Since the birth of radio broadcasting, the dominant role has been played by amplitude-modulated broadcasting—sometimes called double-sideband broadcasting. DSB broadcasting produces two sidebands, when in theory one would suffice. DSB broadcasting is not only wasteful in power, it also wastes frequency spectrum; a given portion of the frequency spectrum could accommodate twice as many radio stations if one sideband could be dispensed with.

Better transmission efficiency has been the goal of successive generations of engineers. To this end the possibility of applying SSB broadcasting to radio has been suggested in the past as a means of alleviating the overcrowded broadcast bands of the frequency spectrum. In 1979 the World Administrative Radio Conference (WARC) passed a resolution to improve the utilisation of the spectrum by the future application of SSB broadcasting. There was a degree of reluctance on the part of broadcasting authorities to welcome this resolution. Nevertheless, since its introduction was inevitable at some time in the future transmitter manufacturers began working on the problem.

At the 1987 WARC, things moved a stage further in the recognition of the problems that confronted the broadcasters. Few of the transmitters already in service could be adapted for SSB. Nevertheless, the International Telecommunications Union insisted that it wished to see SSB broadcasting, and called upon manufacturers to commence development by 1990. Given the extent of the problems, it accepted that widespread SSB broadcasting could not be a reality before the year 2015, at which the broadcasters heaved a sigh of relief.

The problems are of a technical and political nature. SSB broadcasting is not impossible—it has been in use on HF communications since 1947, and even from 1935 some PTTs, such as the British Post Office, had been using it across the Atlantic on radio telephony circuits. Progress in communications was swift, and by the 1960s all commercial authorities, including civil aviation, had made the switch to SSB.

Radio broadcasting, on the other hand, had developed its transmitter technology on different lines. Sound broadcasting is performed by high-level modulation, because of its higher overall efficiency compared with low-level modulation followed by a linear amplifier (as used in communications transmitters). This created a situation worldwide where nearly all radio stations were incapable of being adapted for SSB broadcasting. The only obvious solution was to replace all the high-power transmitters already in service throughout the world, of which it is estimated that there are nearly 2000.

Then there was the political problem. The ordinary AM receiver is quite useless for receiving an SSB signal. Any broadcasting station switching to SSB would therefore lose a large proportion of its listening audience at a stroke. Thus the introduction of SSB broadcasting on a worldwide basis would mean that most existing radio receivers would become useless. Since international propaganda broadcasting competes for its listening audience, it is easy to see why broadcasters should be reluctant to switch to SSB broadcasting. Many international broadcasting authorities who command huge listening audiences (the BBC World Service is one, with 120 million listeners) would no doubt like SSB to be shelved altogether.

To overcome this problem of alienating listening audiences, WARC in 1989 moved to adopting 'compatible' SSB. This would allow the most progressive broadcasters to make the switch to SSB, in such a way that they would remain capable of being heard on a conventional AM receiver. This solution seemed to solve all the problems in that it would allow those forward looking broadcasting authorities to adopt SSB as soon as was practical.

By 1989 Germany, Austria and Norway had begun to implement compatible SSB, and by so doing had pioneered the way. It is significant that VOA, the BBC World Service and Radio Moscow made no move to implement SSB broadcasting. These three are the political broadcasters who command big audiences and wanted to keep them. Since the targets of their broadcasts are listeners with cheap AM receivers in Eastern Europe, Asia and the Middle East, there would be nothing to gain and a lot to lose.

For manufacturers of radio transmitters, far-sightedness is an essential survival trait. It takes ten years to develop a new generation of high-power transmitters, and when a transmitter goes into service the broadcaster reckons on a service life of anything from 25 to 50 years. In 1982 AEG-Telefunken introduced its Pantel range of transmitters, and in anticipation of SSB broadcasting the company made sure that it could produce transmitters for either DSB or SSB operation.

Within a few years, Brown Boveri (now ABB) and Thomson-CSF produced their versions, all suitable for SSB operation. Marconi, however, produced a transmitter that, because of its unique parallel design of pulse-division modulator, would be very difficult to adapt. In 1989 it changed the design to a series modulator which, the company claims, is now suitable for SSB. About this time, Dallas-based Continental Electronics produced its 420 B transmitter, which has SSB facility fitted.

Thus there are now five major manufacturers that have responded to the demand for SSB transmitters. It is reassuring that many international broadcasters have taken the opportunity to re-equip with high-power broadcast transmitters that are not only capable of high efficiency PDM operation (or PSM, in the case of ABB), but are also capable of providing conventional DSB operation, compatible AM and full SSB. In the next five years, many more will have re-equipped in preparation for SSB broadcasting. The next stage in the migration to SSB must be the development of AM receivers suitable for SSB at a reasonable price. At present about 20 such models are on the market, all of which are professional or semi-professional devices selling at prices which are far too expensive for the domestic market. Although there is a growing market for such high-performance radio receivers, if SSB listening is going to become popular then it is up to the receiver manufacturers to develop domestic versions.

There are a number of technical hurdles: SSB receivers must have a much better performance, and frequency stability must be very much higher than in a cheap domestic receiver. This means that such radios will always cost more than the ordinary AM receiver. Eventually the problem will be solved by mass production — the market is potentially huge. In many areas of the world, the ubiquitous MW/SW receiver is the main means of keeping in touch with current affairs.

There is one other problem: the increased skill that is needed to tune into an SSB signal. Any child or uneducated peasant can tune an AM receiver, but it takes care and experience to tune to an SSB signal, and if the receiver is not tuned accurately then the signal may become unintelligible. At present, listeners to compatible AM broadcasts (SSB signals that can be demodulated on an AM receiver) tend to be keen listeners, with sophisticated SW receivers. Thus even the lowest-end mass-market SSB receiver will need sophisticated computer-aided push-button tuning.

Until these problems of easy-to-tune, cheap SW receivers are solved, without sacrificing performance, the broadcasting authorities will continue to delay the switch to SSB broadcasting as long as possible.

Chapter 21

Broadcasting from the Federal Republic of Germany

Germany has more broadcasting authorities than any other country in Europe. There are a number of reasons for this. Not only is it the largest single nation in western Europe, but it is also a federated structure of several states. Its unique relationship with the USA, the UK and France, as the vanquished nation of war, gave the victors some occupation rights, which still exist today. To America, these rights meant that Germany became the cockpit for East-West relations, so that as early as 1948 the USA was planning to build propaganda stations in Germany.

The federal states of West Germany itself boast the following broadcasting stations (or rather authorities, since each may have many different stations): Bayerischer Rundfunk, Hessischer, Norddeutscher, Radio Bremen, Saarlandischer, Suddeutscher, Sudwest and Westdeutscher Rundfunk.

Many of these authorities have powerful stations on AM and VHF-FM. On top of these, the republic has its state broadcasting service, Deutschlandfunk, which has several high-power stations, the highest being of 700 kW. Deutschlandfunk is an information broadcasting service, which also caters for foreign nationals, for whom it broadcasts in many European languages.

As well as the public service broadcasters, there are many commercial radio stations and several foreign stations on German soil, controlled by the occupying powers of France, the UK and USA. There are also a few foreign propaganda broadcasting stations, which have existed since 1950: VOA, Radio Free Europe and Radio Liberty. Berlin is a special case: West Berlin was the showpiece of the West, and for it the Americans created the broadcasting service Rundfunk im Amerikanischen Sektor, RIAS.

The federal German state also operates its own information foreign broadcasting service, Deutsche Welle, which has broadcasting stations in Germany and elsewhere: Julich and Wertachtal in Germany, plus stations in Ruanda, Portugal, Malta, Antigua, Montserrat and Trincomalee. Deutsche Welle is one of the largest international broadcasters in the world, broadcasting regularly in 34 different languages. Its nerve centre is a SW station in Bavaria.

Wertachtal is the highest powered shortwave station in Europe, built shortly before the 1972 Munich Olympics. Before that time Julich had been the most powerful, with nine 100 kW SW transmitters carrying the voice of Germany to the world. Wertachtal was first conceived as early as 1962, when the fully automated station had not been realised—which emphasises the need for long-range planning in international broadcasting.

The many environmental problems associated with high power meant a long and exhaustive search for the ideal site. Over 50 sites were inspected, and some evaluated and abandoned, before Deutsche Bundespost was offered Wertachtal, south of Augsburg. Detailed planning began in 1969, and by 12 June 1972 the first four 500 kW transmitters (AEG S2500) went into service. At the outset the station was planned to take a full 12 transmitters and associated antennas, but even this figure was an underestimate: by the end of 1989 the number had risen to 15, nine used by Deutsche Welle and four by VOA, leaving two spare transmitters.

Even this complement does not give enough scope for future expansion: the federal government sees Deutsche Welle as needing 20 transmitters. The staff at Wertachtal are well pleased with the AEG transmitters; one thing you will not see at the station is a collection of different manufacturers' transmitters. The Germans have every reason to be proud of German technology.

Wertachtal's antenna system is one of the largest in the world, with 76 curtain arrays and one rotatable curtain. The layout is unique. In plan it consists of three bent radial arms symmetrically disposed. Twenty-five steel lattice towers support the four-by-four stacked linear arrays, each pair of towers carrying a horizontal and vertical reflector screen of wires. Either side of the screen are the curtain

Figure 21.1 *Wertachtal SW station, Bavaria, Germany*

arrays, providing the capability to fire in the reverse direction.

As well as this reversibility of the antennas, each curtain array is designed such that the direction of fire can be slewed electrically by as much as 30° in azimuth bearing from the main direction of the lobe; this electronic slewing is achieved by advancing or retarding the phase angle of the elements fed to each antenna. As might be expected, the controlling mechanism for the slewing is itself a very advanced technology. Commands for the control of the switches are issued by a process computer. The line lengths of the phasers used as slewing switches are designed to be infinitely variable, but to simplify the system they have only five positions within the total slewing range of ±30°, each selection fixed by remote control. Thus, each antenna requires five different commands.

Wertachtal is also impressive for its RF transmission line systems. The design of feeder systems to carry transmitter powers of 500 kW carrier power is no light task, made all the more difficult by the need to cater for 100% modulation—and even overmodulation peaks, which impose considerable electrical stresses on the system. Flashovers on a transmission line, once originated, can be self-perpetuating until RF power is removed, so it is important to design transmission systems for AM transmitters with a very considerable safety factor.

From the outset, Wertachtal antenna site was planned around the use of coaxial rather than open wire feeders. The complete system comprises some 53 km of coaxial cable, with 450 couplings, and the whole installation is airtight and pressurised with dry air, making it impervious to climatic conditions. Since the time of commissioning Wertachtal has not had a single incident of a cable fault. This area of the lower slopes of the Bavarian Alps is no California—environmental conditions are harsh, ranging from gale-force winds and driving rain, to sub-zero temperatures with heavy snowfalls and severe icing. Thus the planners of this transmitting station were faced with a formidable task calling for ingenuity, considerable experience and a great deal of money. That the federal government budgeted the expenditure for Wertachtal transmitting station, the largest of its kind in Europe—exceeding the capability of the BBC SW station at Rampisham, England, or that of RFI's SW station at Issoudoun—emphasises the importance attached by Germany to information broadcasting. With its fifteen 500 kW SW transmitters there are very few places in the world that Wertachtal cannot reach; its transmitters carry the voice of Germany—and that of America—to all corners of the world.

The design and construction of a fully automated, menu driven, high-power SW broadcasting station of the size of Wertachtal calls for skills of a high order in many disciplines. European engineers historically, partly from necessity, are trained in multi-disciplinary related technologies, which explains why the giant companies of Europe excel in such projects. Wertachtal was a joint European project, in which AEG-Telefunken supplied the transmitters, associated equipment and control systems, while ABB designed and engineered the antennas and their control systems.

21.1 RIAS

Rundfunk im Amerikanischen Sektor is a US-funded broadcasting station in the American sector of Berlin. Because of the political situation that has existed since

Figure 21.2 *500 kW shortwave transmitter showing PA tube TH558 and tuning mechanism*

the Second World War, Berlin has more broadcasting stations than any other city in the world. There are German radio stations, stations in the British sector, the French sector and the American sector. On top of all these the BBC World Service has a low-power relay station.

Of this plethora of radio stations, pumping out entertainment of various kinds in many languages on all wavebands, RIAS is the most interesting. As a transmitter station, its history dates back to 1946. The smoke and flames of the war had long since disappeared but the city was in ruins; Berliners were for the most part homeless and short of fuel, water and electricity. On top of all these problems, the winter of 1946 was one of the harshest in two decades. Conditions were worsening with every passing week, and the situation was aggravated by the fact that the city was isolated from the West.

Added to these problems was the lack of adequate communications. Telephone and telegraph lines were disrupted and the broadcasting stations were out of action. German ingenuity came up with a solution: to string telephone wires around the city suburbs. This hastily improvised method would permit messages and instructions to be passed to Berliners through loudspeakers—one of the first examples of wired radio broadcasting. However, there remained the problem of a signal source. For this purpose the German engineers modified a 50 kW LW transmitter to operate on a frequency of 50 kHz. At strategic places around the city loudspeakers were connected to the telephone lines.

The modified radio transmitter for this hook-up was none other than the mobile radio station used in the North African campaign, famous for its nightly broadcasts of the haunting tune of Lilli Marlene. It was a fitting end to this famous radio station which had tugged at the hearts of German and British soldiers alike in the desert of 1942–43. The Lilli Marlene transmitter still occupies a deep affection in the hearts of all who remember it, and particularly Berliners, not only for the war years but also for the part it played during the blockade of Berlin. In 1987 the message was passed on to the younger generation, when RIAS staged a pop concert attended by 20 000 young Berliners, to celebrate the 40th anniversary of the blockade and the 750th anniversary of the founding of Berlin.

From such humble beginnings, RIAS acquired its first high-power transmitter in 1949, after the lifting of the blockade. This first voice of Berlin was a 100 kW MW transmitter designed and built by AEG, the same company that had built the Lilli Marlene radio stations. When I visited RIAS in 1988, the Lilli Marlene transmitter occupied pride of place in the museum of transmitter items at Britzer Damm. It comprised 12 four-wheeled trucks, each carrying part of the radio station: one truck carried the generator, another the radio transmitter and so on.

In 1953 RIAS acquired a second transmitter: a 300 kW MW transmitter, which now radiates a 24 hour transmission on a frequency of 990 kHz, with the older 100 kW radiating on 855 kHz. There are two MW antennas, one half-wave dipole, one cross-polarised type developed by AEG to give optimum long-distance coverage to East Germany during the hours of darkness.

In 1983 RIAS acquired its third transmitter, another AEG model with an output of 100 kW for broadcasting on short waves. This transmitter broadcasts on a single assigned frequency of 6005 kHz. It is connected to a folded dipole antenna, which is broadly omnidirectional in radiation. As well as its coverage on medium and short waves, RIAS also broadcasts on VHF frequencies 94.3 and 89.6 MHz. With output powers of 20 and 10 kW there is no doubt whatsoever that RIAS programmes were intended to be heard in East Berlin and beyond into East Germany.

The only waveband that RIAS does not broadcast on is long wave. This is a legacy of history. One of the penalties that Germany paid for losing the war was that it was deprived of its broadcasting facilities in the MW and LW bands—just as, after the First World War, Germany was deprived of all its transoceanic and intercontinental telegraph cables. In both instances, the aim was to deprive the nation of its international means of communication.

In the place of medium and long wave, Germany was allowed to use VHF—which is only suitable for short range use and has no strategic value. As a result

Figure 21.3 *RIAS Berlin control room at Britzer Damm SW station, 1989*

of this restriction, Germany pioneered the art and science of VHF FM broadcasting; the irony for the victorious Allies of the war is that Germany now leads the world in VHF technology, through such companies as Rhode & Schwarz and AEG Telefunken.

As well as being an operational broadcasting station RIAS is a museum of transmitter equipment – much of it still in working condition, though long since relegated to standby duty. During my visit I saw the 100 kW MW transmitter of 1949 vintage switched on to a heater condition, and after one minute it was brought up to full carrier power. This transmitter had done over 100 000 hours service.

21.2 The covert history of RIAS

My visit to RIAS was quite unplanned. Through the generosity of AEG, my wife and I travelled to Berlin in December 1988, having been invited to the works at Sickingenstrasse, Berlin to see some examples of AEG transmitters. It was an unexpected bonus to be invited to visit the high-power transmitter site at Britzer Damm, the nerve centre of RIAS.

At that time I had not researched German postwar broadcasting history nor had I researched the early history of the US Central Intelligence Agency and its involvement with covert broadcasting. My reception at Britzer Damm could not have been better, and I was immediately struck with two things. One was the obvious importance of the transmitting station, with its capability of broadcasting 24 hours a day on MW, SW and VHF-FM bands. The other was the close bond

Figure 21.4 *Transmitting tube museum at RIAS Berlin Britzer Damm SW station, 1989*

of trust and friendship between the staff of RIAS and the senior management of AEG-Telefunken.

The visitors book made impressive reading: it contained the signatures of VIPs up to and including Ronald and Nancy Reagan, who had been there nine months earlier to celebrate the 750th anniversary of the City of Berlin. The 94 page book issued to commemorate this event gave details and pictures of an open-air concert staged at the Brandenberg Park attended by 20 000 Berliners, who were entertained by a galaxy of pop stars, including the likes of David Bowie, Stevie Wonder and Tina Turner. It was one of the largest rock concerts ever staged, with a staggering billboard.

This concert went on for days and nights. Significantly, it was held close to the Brandenberg Gate in the Berlin Wall. Powerful loudspeakers carried the sounds of the superstars into East Berlin, where thousands of young *Ossis* had gathered. President Reagan delivered one of his most famous speeches from a rostrum that had been set up close to the Wall, on 12 June 1987.

All this was staged and broadcast by RIAS, using its own symphony orchestras. Few broadcasting stations in the world own a philharmonic orchestra like that of RIAS. The youngsters in East Berlin went wild with excitement at hearing, while being unable to see what their peers in the Western sector were enjoying. When David Bowie, Phil Collins, Dave Stewart and Tina Turner performed there was rioting in East Berlin.

There can be no doubt that this was what the concert had been staged for; even the special commemorative book issued by RIAS contains some pictures of headlines about rioting and police arrests on the other side of the Berlin Wall.

It was several months later, when researching the history of the CIA, that I discovered that the setting up of RIAS had been carried out and funded by CIA during the ten year reign of Allen Dulles. This was by all accounts the most infamous period in its entire history. As director of the CIA, Dulles controlled an organisation far richer and more powerful than anything achieved by the Abwehr in Nazi Germany, the KGB in the Soviet Union or the SIS in Britain. His budget in 1949–50 was $97 million—a staggering amount for that period.

A small part of this budget went on the initial setting up of RIAS, for in that year it took delivery of its first high power SW transmitter. During that period of the Cold War, which was fast approaching its most dangerous phase, Berlin was the cockpit from which America waged its propaganda war against the Soviet empire. RIAS was one of its weapons.

No doubt the setting up of a propaganda broadcasting station only a mile or so from East German territory disturbed the East Germans and the Soviets, for it signalled the beginning of a broadcasting race, with both sides expanding their output. The first new station in East Berlin was Radio Berlin International. It began broadcasting in 1955 in English and French languages, and Swedish, Danish, Arabic and German followed soon afterwards. Its programmes were extended to include Italian, Spanish, Swahili and Portuguese in 1964, and Hindi in 1967. Today RBI broadcasts some 75 hours of programmes a day in 11 languages, mostly on short wave but also on medium wave to Europe, Africa, the Middle East, Asia, Australia and North and South America.

As with most international broadcasters, RBI has always welcomed listeners' reports on its broadcasts, and in common with many others it provides a blank reception report for the convenience of its listeners. After the column asking for reception details, there is a column for type of interfering signal. It would appear that RBI has experienced, or expects to experience, some form of radio jamming of its broadcasts. The major Western broadcasters—the BBC, VOA and DW—have always condemned jamming, which leaves open to conjecture the question of who might have been interfering with RBI broadcasts.

Chapter 22

Scandinavian broadcasting

Scandinavia has never sought a high profile in information broadcasting. The league table of international broadcasters issued by International Broadcasting & Audience Research includes only one Scandinavian country – Sweden – which it places seventh from the bottom. Norway, Denmark and Finland do not even get a mention.

In fact these countries have a low output in a few languages. In Sweden, Sveriges Riksradio broadcasts in English, French, German, Portuguese, Russian, Spanish and Swedish. In Norway, Norsk Rikskringasting broadcasts in six European languages. Finland's Yleisradio broadcasts in four European languages, and Danmarke Radio broadcasts in Danish and English. There are two other countries in Western Europe that roughly fall into the same pattern: the Netherlands and Switzerland. The Netherlands broadcasts in seven languages, including Arabic and Indonesian, while Switzerland broadcasts in six languages, all European.

All of these countries have in common that they have no wish to enter the arena of world politics, nor do they have pretensions about exercising a global influence. None of these countries took a major role in the Cold War, nor were their radio broadcasts jammed by the Soviets. Although it would be easy to conclude that these countries are not seriously interested in international broadcasting, this would be incorrect. They might be better regarded as broadcasters of quality rather than quantity, as is evident from a study of the broadcasting standards and technology.

Sweden acquired 500 kW state-of-the-art transmitters in 1973, and Norway followed in 1976. Both were of Thomson manufacture. Norway acquired more of the same technology in 1982, as did the Netherlands in 1983. In these cases the transmitters were supplied by AEG-Telefunken. The Norwegian SW station at Kvitsoy, when completed by AEG-Telefunken in 1983, had four 500 kW PDM transmitters, which were far ahead of anything possessed by the BBC. Two years later the BBC attained parity when it bought four of the same transmitters (the S-4005 from AEG). Kvitsoy is one of the few SW stations to be equipped with a 500 kW rotatable high-gain curtain array. This type of antenna is very expensive and very versatile, with the ability to radiate on any bearing around 360°. Because of this it can replace many fixed curtains.

Sweden, Norway and the Netherlands each have four 500 kW SW transmitters and many more of lower power. But the latest pace-setter in the Nordic world is Finland.

Figure 22.1 *500 kW shortwave transmitter S4005 installed at Flevoland, the Netherlands*

Pori

Pori is the SW station of the Finnish Broadcasting Company, Oy Yleisradio Ab (YLE). YLE began its history in the 1920s which makes it one of the pioneers of radio broadcasting. From national broadcasting it moved up to international broadcasting in the 1930s, beginning with three 15 kW transmitters, and after the war (1948) it acquired a 100 kW transmitter all from Brown Boveri. Quite evidently YLE got good service from these transmitters, because it did not get another refit for a further 20 years when it expanded its SW broadcasting operations with the acquisition of a 250 kW transmitter together with a similar-rated log-periodic antenna. By the late 1970s, however, the station was becoming obsolete in other ways and it was also being encroached upon from a residential area. The time was ripe to move.

After an exhaustive search to find a site with the necessary electrical qualities for efficient propagation, a suitable area was found close to the sea, about 15 km from Pori. Having found the ideal site it was decided to use it for both national and international broadcasting. The planning of a new site is not something that can be done in a matter of months. In the case of this new station the planners had the advantage of a virgin site. The new transmitting station was designed from the ground upwards, and was able to benefit from the latest technology on the market.

Planners frequently experience considerable problems in introducing 500 kW transmitters to existing, older stations—inadequate power supply, equipment congestion and inability of the transmission lines to handle the super power all generate stresses on existing systems. Pori did not suffer from the kind of problems that afflicted the BBC when it introduced its first 500 kW into service.

The new transmitting station at Pori was the first multi-frequency station: it has MW and SW transmitters within one complex. This is the latest trend in transmitting stations, made possible in recent years with the advent of computer-aided design using modelling techniques.

A feature of the new three-story building is the meticulous attention paid to controlling and minimising electromagnetic radiation. Each room is screened, and each of the high power transmitters is housed in its own fireproof cubicle to contain any outbreak of fire. This is a far cry from the old method of installing rows of transmitters in one large open building. Despite all the technological advances that have been made over the years, the three ingredients of wind, water and high voltage make a lethal cocktail in the event of fire. The computer is increasingly relied upon for a variety of applications, so the computer control room is completely shielded by a Faraday cage, and all signal inputs and outputs are shielded.

Pori now has five transmitters. Three of these are new 500 kW SW type, fitted with PSM, suitable for DSB or SSB operation. Dynamic carrier control (DCC) is also fitted. The fourth is a new 600 kW MW transmitter, also fitted with PSM and DCC (but without SSB, which is not a feature of MW broadcasting). The fifth transmitter is the 250 kW SW transmitter moved from the old site, and it has been upgraded for DCC operation.

The site now has a total of 12 antennas. Ten of these are high-gain curtain arrays (HRS 4/4/1, with four dipoles stacked in a vertical and horizontal arrangement and a reflecting wire curtain). Making up the complement of SW antennas is a log periodic antenna, covering the entire SW frequency range from 5.95 to 26.6 MHz. This log periodic antenna is rotatable, covering the full 360° of azimuth, and serves the same purpose as the rotatable curtain array described earlier. The MW antenna is a dedicated item for the 600 kW transmitter with its associated antenna matching unit. Its feeder is a 6⅛ in coaxial cable.

The Finnish policy has been to acquire an ultra-modern integrated transmitting

Figure 22.2 *The Pori SW station (Finnish Broadcasting Company) under snow*

station for multi-frequency, continuous broadcasting. This station can run unattended for long periods. It is menu fed by the computer, which controls every function from change of mode (DSB/SSB/DCC), to change of frequency in all four SW transmitters, selection of antennas and azimuth bearings for directing SW broadcasts.

In its efforts to acquire the best technology, YLE procured items from different contractors. Antennas came from TCI of America, radio masts and support structures from Finland's Imatran Voima, and ABB supplied the radio-frequency electronics: transmitters, RF subsystems and a two-part antenna-switching matrix. Control and monitoring system, supplied by the Finnish company Altim Control, consists of eight computers. The station allows three types of control: full remote from Helsinki, remote control from the station console, and manual control from the transmitters.

When the station was commissioned in the summer of 1988 it was the most modern in Europe. Since then it has operated without serious fault or interruption to service. Finland, to its credit, is one of the first countries in the world to use compatible SSB in its broadcasts. The new station is a far cry from the old SW station designed in the 1930s, which was described to me as 'The world's only functioning radio transmitter museum'. This is the nicest compliment to pay to transmitters that are taken out of service not for unreliability, but simply because they cannot meet the operational requirements of single-sideband broadcasting in the year 2015.

Figure 22.3 *The control room at the Pori SW station*

Chapter 23

LW and MW international broadcasting

The first radio broadcasts were heard on the long and medium wavebands over 70 years ago, when the power output of the transmitter was measured in watts. When better tubes became available, it was possible to design transmitters for AM broadcasting with carrier powers of a few kilowatts. By the mid 1930s the average carrier power of a broadcast transmitter was 50 kW.

Since those early days significant advances have been made in transmitter design, both in terms of electrical conversion efficiency and output power. These advances have come about entirely as a result of improvements in components such as vacuum capacitors, solid-state rectifiers and transmitter tubes. Indicative of the technological strides that have been made in tube development is the TH-539, a power grid tetrode that by virtue of hypervapotron cooling, can dissipate 1 MW in LW/MW service, and is therefore capable of providing a carrier power output of 1.25–1.5 MW.

Typical AM broadcasting transmitters now in service on the long and medium wavebands have power outputs of 1 MW. Sometimes a number of such units are combined in a summing arrangement to produce several megawatts of power. The spur to these developments has been the shift in emphasis in broadcasting applications, from national to international broadcasting.

From the early days of broadcasting the MW and LW bands were perceived as the perfect medium for regional and national entertainment broadcasting. Long waves are a particularly stable mode for radio broadcasting, without fading or seasonal variations, and not subject to the problems caused by ionospheric reflection. The ionosphere makes long-distance communication possible on short waves, but it also affects MW broadcasting after the hours of darkness.

Broadcasting by long waves is stable, but it is also very expensive: to provide a reasonable signal strength at a distance of several hundred miles requires a carrier power of 1 MW or more. Nevertheless, since information broadcasting has become an integral part of foreign policy, the cost has become a secondary consideration.

Medium waves exhibit characteristics that for many decades were regarded as a disadvantage or handicap. The so-called 'night-time effect,' or, more properly, 'selective fading' is a phenomenon that is at its worst at a certain distance from the radio transmitter when the waves following a low path along the ground and those refracted across the sky are of equal amplitude but in opposite phase, causing destructive interference and loss of signal. Up to this critical distance the ground wave prevails, although fading and unsatisfactory reception occur. To combat this, the service area of a national radio station was always reckoned as being within the distance of the ground wave, generally up to 50–125 miles depending on the power output from the transmitter.

The growth of international information broadcasting since the 1950s prompted the realisation that the MW night-time effect could be used to some advantage. However, by tradition the MW and LW bands were accepted as the province of national and regional broadcasters, leaving the short waves to international broadcasters.

If we draw a veil over the Allies' activities with super-power transmitters in the Second World War in terms of subversive broadcasting, then the first dissentient to the accepted rule came in 1953 when American high-power MW transmitters were installed in a ship in the Mediterranean for propaganda warfare. That same year the first 1 MW transmitter came into service: it was installed in Munich, the programme feed came from Washington, and it was tuned to the same frequency as Radio Moscow—the licensee of the frequency 173 kHz—using two to three times the power of Moscow.

From that time on, the accepted standards of international broadcasting had gone forever. AM super power had demonstrated its awesome potential and there was no turning back. The company that built the first megawatt transmitter was Continental Electronics in Dallas, Texas. In fact, this 'megawatt' transmitter consisted of two 500 kW units coupled. Nevertheless, a few years later the company built a true 1000 kW unit. The logical evolution from this achievement was the building of a 2 MW transmitter, by combining two 1000 kW units.

23.1 Changing trends

SW radio broadcasting is still the only medium with the ability to reach out to a listening audience thousands of kilometres distant with the assurance that reception cannot be censored or prevented. Some parts of the world also use short waves for domestic broadcasting; generally, these are the large land masses such as China and Russia. This means large numbers of people in possession of AM receivers with a SW facility. For these reasons, short waves will continue to be the main vehicle for propaganda broadcasting.

With the introduction of television and frequency-modulated sound broadcasting on VHF, there has been a significant change in habits in Europe. From what was a popular pastime some 50 years ago, listening to SW broadcasts is now practised by a tiny minority. The latest available statistics show that only 25 out of every 1000 people in Western Europe take the trouble to listen to SW radio; television has become by far the most popular form of entertainment, with FM radio running second.

Manufacturers of domestic radio receivers, recognising the trend, have in many cases ceased to build an SW capability into their designs. There are many excellent professional and high-grade, all-waveband receivers on the market, but some of these are very expensive. By contrast, the cheap domestic radio receiver capable of receiving AM broadcasts on LW and MW wavebands is still in common use, not only in Europe but throughout the world. Major international broadcasters have all recognised the declining popularity of SW listening in certain parts of the world.

23.2 Time-shared broadcasting

High-power AM transmitters have become increasingly complex in design and performance, and this trend towards sophisticated technology has been accompanied by quite dramatic increases in output power. What was once regarded as super power has become the norm, such that 1 MW and even 2 MW carrier power transmitters are in regular use in several countries.

There have been other changes in AM transmitters. From the old fixed-frequency transmitter for broadcasting on LW and MW bands, even the largest modern transmitters can be retuned to alternative frequencies in a matter of minutes. The modern transmitter can operate round the clock, shutting down only for maintenance checks. (Of course, not all broadcasting authorities transmit round the clock—many shut down in the early hours.)

Figure 23.1 *Radio Monte Carlo, Roumoules, multi-megawatt LW/MW transmitting station*

Time-shared broadcasting is the answer to the increasing costs, not only in initial capital outlay but in daily running costs, which in the case of a 1 MW transmitter can be equal to the electricity consumed by a medium sized city. Making better use of transmission technology, and the use of ionospheric wave broadcasting, a high-power MW transmitter can provide nationwide coverage, with a primary (ground wave) service area to a radius of over 1000 km. At night, switching to a higher frequency, the same transmitter taking advantage of sky-wave propagation can reach international audiences in other countries. When this is used in conjunction with a directional antenna, a dramatic improvement in the service range becomes possible, as the gain can be as high as 14 dBi. The range of a megawatt transmitter can be extended to over 3000 km. For example, the new transmitter complex at Roumoules, sited on a plateau of Les Hautes Alpes, can blanket the whole of Europe at night, well into Russia almost to the Urals.

Chapter 24

Jamming on the short waves

Jamming is usually presented as a straightforward moral issue, but as in all moral issues there is in fact a large 'grey area'. The views of both sides—those whose broadcasts are jammed, and the jammers—need to be weighed against the evidence from both sides. Although jamming of radio broadcasts is a violation of ITU regulations, and was condemned by the Helsinki Act of 1975 which recognises the right to 'free flow of information,' this can be counterbalanced by the 1936 and 1957 Geneva Resolutions condemning broadcasts likely to 'incite the population of any territory, to acts incompatible with internal order.' With such an obvious conflict it is perhaps not surprising that radio jamming has gone on for as long as it has, with both sides claiming to be the injured party.

24.1 Regulating SW broadcasting

Within the HF spectrum, from 3 to 30 MHz, there are nine bands of frequencies that are reserved for broadcast services. As a result of growth in the past three decades, more than 150 countries broadcast in these shortwave bands. Countries assign frequencies to their stations in accordance with allocations drawn up under the auspices of the International Telecommunications Union (ITU) in Geneva. This body, a specialised agency of the United Nations, is responsible for regulating all forms of international telecommunications.

Of all the frequency spectrum allocated for communications and radio broadcasting, it is the HF spectrum, 3–30 MHz, that is the most difficult to administer and regulate. One reason is the unique nature of this particular frequency spectrum. Unlike VHF and UHF, which are effectively line-of-sight communications, HF knows no such limitation. Propagation may be by ground wave or by ionospheric reflection, more commonly referred to as sky wave. As a result, the radio signal knows no boundaries; it can span oceans and continents with ease. More often than not, the signal at a distance of 4000 km may be stronger than at 100 km.

The second reason for the difficulty of regulating SW broadcasting is the numbers of transmitters sharing this spectrum. Several nations have more than the one broadcasting authority, and each authority may broadcast in all the SW bands at any one time, filling a very large number of frequencies. It is currently estimated there are over 2000 medium to high-power stations in use, and there is a definite trend towards phasing out the medium powered stations and replacing these with stations of 300–500 kW carrier power.

Most of these stations are foreign-service information broadcasters. Information broadcasting can have a number of objectives, ranging from projecting a nation's culture and presenting a sympathetic image to the rest of the world, to a type of foreign policy bent on exploiting political happening in other parts of the world.

24.2 Postwar Soviet jamming

In February 1948, when relations between East and West were deteriorating in a process that was to culminate in the Berlin blockade, the Soviet Union began jamming Russian language broadcasts from VOA's transmissions. In 1950 most Eastern Bloc countries followed suit, and by the end of 1951, virtually all local-language broadcasts beamed to Eastern Europe were being jammed.

In 1956, Poland stopped jamming following the 1956 Poznan riots, during which rebel workers attacked, among other facilities, local jamming stations. Intermittent jamming returned to Polish airwaves around 1959, but it is not clear whether this emanated from within or outside Poland's borders.

The first lessening of jamming occurred in September 1959 during Soviet Premier Nikita Khrushchev's visit to the USA. Some informal bilateral discussions led to the end of most of the jamming against VOA and the BBC except for selected news items and commentaries. However, selective jamming resumed after Khrushchev returned to the Soviet Union.

In June 1963, following President Kennedy's famous American University speech calling for US-Soviet negotiations to end the Cold War, the Soviets stopped jamming VOA broadcasts for the first time in 15 years. Soviet jamming of most other Western broadcasters also stopped, but jamming of Radio Liberty continued. In July 1963 Romania ceased all jamming, followed by Hungary in February 1964. In April 1964 Czechoslovakia also discontinued its jamming of VOA and the BBC, but not Radio Free Europe. This relief was short-lived, as massive Soviet jamming of the VOA, BBC and Deutsche Welle resumed in August, when Soviet troops and tanks invaded Czechoslovakia. Broadcasts in Czech, Slovak and several languages spoken in the USSR were affected.

In 1970, following serious worker riots on Poland's Baltic coast, Poland resumed its jamming of Radio Free Europe. In 1973, on the eve of the initial meetings in Helsinki of the Conference on Security & Co-operation in Europe (CSCE), jamming of VOA, BBC and Deutsche Welle broadcasts in Russian and minority languages of the USSR ended. Jamming of Radio Liberty continued. Following the Egypt-Israeli War of 1973, the Soviets intensified jamming of Soviet language broadcasts from Kol Israel as well as the broadcasts in Hebrew and Yiddish.

By 1978, almost all Soviet and East European jamming had ceased except for that directed against Radio Free Europe and Radio Liberty. In August 1980, however, following Polish government concessions to the Solidarity movement, the USSR resumed massive jamming of VOA, BBC and Deutsche Welle broadcasts to the Soviet Union. In December 1981 jamming of Western Polish language broadcasts resumed, presumably from jammer transmitters in the USSR, East

Germany and Czechoslovakia. Bulgaria joined in the jamming about 10 months later. Jamming of Polish language broadcasts intensified in 1982 following the imposition of martial law in Poland and the banning of Solidarity. In 1982 jamming was initiated against Western broadcasts to Afghanistan.

In October 1986, the Soviets stopped jamming Chinese government broadcasts in Russian; in October 1987, jamming of BBC broadcasts in Russian ceased.

The history of Soviet and East European jamming shows that it has varied in accordance with East-West relations, tending to increase during periods marked by internal and external tension. An analysis of the above chronology of jamming, extracted from a publication issued by the United States Information Agency (USIA), confirms that jamming is characterised by a series of high pressure points followed by an easing of tension, but accompanied at all times by a constant underlying degree of mistrust on both sides.

The chronology of events given by VOA, although true, presents a selective picture. It fails to mention that out of the 150 countries that use the short waves for information broadcasting, only four have been the subject of Soviet jamming operations: the UK, West Germany, Israel and the USA. Such leading broadcasters as Austria's ORF, Canada's RCI, France's RFI, Monaco's TWR, Portugal's RDP, Sweden's SRI, Finland's YLE, the Netherlands' RNI and Japan's NHK have never been subject to Soviet jamming.

Nor can the Soviet jamming of certain broadcasters be attributed to the fact they are broadcasting in Eastern European languages. Many of those broadcasters who were not jammed regularly broadcast programmes in these languages and beamed by short wave to the country concerned. Trans-World Radio, a religious station described later in this book, regularly broadcasts to the Eastern Bloc countries, the Baltic States, and to many different ethnic groups within the 15 constituent Republics of the USSR, yet its broadcasts have not been the subject of jamming.

The key words in a VOA report on Soviet jamming are VOA, BBC, RFE, RL, Deutsche Welle and Kol Israel. These broadcasting agencies work in concert with each other. The BBC carries VOA programmes, DW carries VOA programmes at different times. In some cases the transmitting stations are shared by these broadcasting agencies.

Other key words in the VOA report are 'riots', 'disturbances', 'martial law' and 'attacking installations'. Propaganda broadcasting finds the weak points in a prevailing political situation and exploits them. When broadcasting agencies constantly claim to be the voice of truth, they are giving the truth of the country to which the broadcasts are directed. This is a one-sided truth; at the same time, they may be concealing the truth about what is happening in their own country. The VOA report also does not go into detail on the operations of Radio Free Europe and Radio Liberty, other than to say that these are privately owned radio stations funded from voluntary contributions.

24.3 The Soviet side

The USSR never denied jamming broadcasts from the BBC and VOA. Its stance on the subject was always the same: that a country is entitled to take whatever steps are deemed necessary to jam those radio broadcasts that seek to exploit political situations, or are likely to incite the population to acts incompatible with law and order.

What the USSR did was to set up a network of radio jammers, each of which used one of two methods: jamming by ground wave or by sky wave. In some areas a combination of the two systems was used. Basically, a jamming station is like an ordinary transmitter station—any transmitter may be used as a jammer—so any country has the available means to jam any broadcast it wishes. In general, the difference between a ground station jammer and a sky wave jammer is one of output power.

Jamming by ground wave is efficient in high population density areas where there are known to be large numbers of listeners tuning in to foreign broadcasts. Thus a number of 5 and 10 kW jammers could be installed at strategic intervals around a city, each covering a portion of the HF spectrum. At the point where the ground-wave signal runs out, so does the usefulness of the ground-wave jammer. Ground-wave jammers are unsuitable for sparsely populated regions or for jamming over great distances: the effective range is likely to be up to several kilometres.

Sky-wave jammers, on the other hand, need large amounts of radiated power in conjunction with sophisticated high-gain directive HF antennas favouring low-angle radiation conducive to good sky-wave propagation. To be effective, a sky-wave jamming installation needs to have several high-power (or better still, super-power) transmitters in conjunction with a large number of directional antennas, and a switching matrix that will permit any of the transmitters to be connected to any one of the antennas, which are beamed towards different target areas and countries.

Ideally, the transmitters will be fitted with self-tuning capability, so as to be capable of re-tuning in a few seconds to the frequency of a pilot signal fed into the transmitter. This pilot signal could be the output from a radio receiver of the signal to be jammed. Computer control of the transmitting station is very helpful to successful jamming operations.

Of course, any modern, SW transmitting station, with a dozen 500 kW super transmitters, frequency-agile and computer-controlled, able to retune to any new frequency in a matter of seconds, has all the qualities demanded of a jamming installation. Indeed, the specifications for such stations now being planned by VOA and other broadcasting agencies could have been written with the operational requirement in mind.

24.4 Barrage broadcasting

Naturally, Western broadcasters developed an arsenal of techniques to counter the effects of jamming. From 1949 the BBC and VOA realised that if their propaganda broadcasting was to be at all successful, they had to devise means of negating the effects of Soviet jamming operations.

Measures adopted included:

- mobilisation of all available transmitters
- overmodulating a transmitter
- increasing the average power in the modulation envelope by speech clipping

- saturation, or barrage broadcasting
- programme techniques: repetition, articulation, context.

The first of these may be self-evident. The BBC and VOA, in common with other international broadcasters, broadcast to the world. Thus by closing down some of the broadcasts the transmitter stations are freed to carry anti-Soviet propaganda.

Overmodulating, if carried out continually, will shorten the life of the transmitting tubes, thereby adding to station operating costs. However, it can be a useful measure for special occasions. Speech clipping is a useful device which has the object of increasing the average power in the modulation envelope. In the early days of broadcasting speech clippers were fairly crude in design: they clipped the peaks of modulation, thus increasing average depth of modulation, but they also caused speech distortion and sideband splatter. The speech processors of today are sophisticated in design and performance and make an effective contribution to SW broadcasting.

Saturation broadcasting, or barrage broadcasting, was a technique pioneered by VOA acting in concert with the BBC about the time of the Berlin blockade. The technique involves arranging for a substantial number of high-power transmitters, operating on different frequencies, all to carry the same programme. The reasoning is that some will get through the screen. The more radio transmitters operating at any one time, the greater the probability that the Soviet jammers would not be successful against all the transmitters all of the time. During the Berlin blockade of 1949, and at various times since, the BBC and VOA have been operating as many as 50 transmitters at any one time, all carrying the same message.

Other techniques were to do with programming: improving the quality of articulation, the use of very short sentences, which repeat the same message several times, interspersed with long pauses, were found to be most effective in combating the effects of jamming, enabling the listeners to grasp the all-important context.

The combination of all these measures is costly, to the tune of as much as 100 times the cost of a normal transmission. The fact that broadcasting agencies deemed it necessary to adopt these drastic measures merely serves to emphasise the importance attached to propaganda dissemination by SW transmission. Where world politics are at stake, cost is of no consequence.

Morally, there is another aspect to consider. The HF spectrum is not an infinite resource; the use of as many as 50 different frequencies to transmit the same message is open to the same condemnation as jamming. Both are wasteful of valuable parts of the frequency spectrum. The HF spectrum has always been the most congested of the bands that are used for terrestrial broadcasting.

24.5 The end of Soviet jamming

On Friday 1 January 1988, BBC monitoring staff at Caversham were among the first to observe a significant happening in international relations on the HF spectrum; Soviet jamming of BBC broadcasts in Polish had suddenly stopped, for the first time since martial law had been imposed in 1981.

The cessation of jamming of the BBC broadcasts was unexpected, and threw the BBC World Service into a quandary: was it permanent or only temporary?

If the former, then it called for a drastic reappraisal of programme structure. In the words of a BBC spokesperson, it meant that programmes no longer had to suffer endless repetition to counter the effects of jamming. Secondly, the availability of this extra air time enabled the BBC to put out a more balanced schedule—less emphasis on hard politics, more programmes of entertainment; rock, popular music, classical music and discussions. Apparently this better mix of programmes did not go unnoticed by listeners in Eastern Europe. It was reported that listeners in the USSR dubbed the change as the BBC's own *perestroika*.

1989 marked the implementation of Mikhail Gorbachev's policies of *glasnost* and *perestroika* in some real practical measures. One of these was the dismantling of Soviet jamming installations that had been in place from the early 1950s in many regions of the USSR, particularly in Russia and Byelorussia. On 26 May 1989 the Soviet news agency Tass released the following news item:

> 'The dismantling of jamming devices was started today at Byelorussia's largest jamming facilities situated in the centre of the Byelorussian capital, Minsk. These devices started operating in 1953, the equipment was certainly modernised more than the once. The latest equipment for transmitting in SW broadcasts prepared by Creative Workers Union and trade unions, public organisations, as well as programmes of Byelorussian radio, is to be installed in place of the complexes that are now being dismantled. These facilities will also help improve radio and telephone communication of industrial and transport enterprises.
>
> ' "The decision to stop jamming of broadcasts from abroad is another step forward in the development of *glasnost* and a real result of the Vienna agreement", said Ivan Gritsuk, Minister of Byelorussia Republic of the USSR. The participating countries reached consensus about the need for broad access to all kinds of information.
>
> 'This step is sometimes explained in the West by pressure on our government, a high cost of jamming installations. But these assertions are groundless. In reality, the refusal from jamming, the dismantling and re-equipping of all five jamming facilities in Byelorussia, just as in other regions of the USSR, is connected with a cardinal turn from confrontation to a dialogue in East-West relations. New conditions must prompt all participants in the process of open exchange of information to greater responsibility for the results of their activity. It must be based on the principles of humanism and co-operation and of discarding the stereotypes of the Cold War.'

Chapter 25

'Speaking peace unto nations': BBC World Service

While most international broadcasting agencies take pride in telling the world they are the voice of the state and the instrument of its foreign policies, the BBC steadfastly clings to the statement that it is completely independent of government control. America's foreign broadcasting service is overt in its claim, that it is *the* Voice of America, and much the same can be said for nearly 100 other foreign broadcasters. Not so the BBC. It has successfully traded on the claim that it is an independent broadcaster ever since it first entered the field of international broadcasting in the early 1930s. Few of its dedicated listeners around the world—of which it claims some 120 million—would ever believe anything different. Yet the BBC depends on grants from the Foreign & Commonwealth Office for its continued operations, and its very existence as one of the top five broadcasters in the world.

The BBC World Service has undergone several name changes since it began its Empire broadcasts in 1927, first on an experimental basis and then as a regular feature from 1933. The character and style of its broadcasts have been honed and refined over the years, such that it is now the envy of many of the world's major broadcasters. It has set a world standard by which others are judged. If it were easy to emulate the standard which has been set by the BBC, no doubt it would have formidable competitors; as it is, the BBC World Service remains unique. It has been described by one of its competitors as a subtle and refined tool for influencing minds, based upon Jesuitically refined principles.

The BBC World Service is currently broadcasting more programme hours than at any other time since the 1950s, when the BBC's output to the world was greater than the sum of the outputs from the USA and the USSR. In terms of global coverage its stature today is even greater than it was four decades ago. Anyone examining its relay network, which is strategically located at key points around the world, might be forgiven for believing that Britain had never lost its empire. In the Western Hemisphere it has the Ascension Island relay station, equipped with six 250 kW SW transmitters, and in the Caribbean it has four 250 kW SW transmitters at a station jointly owned between the BBC and Deutsche Welle, radiating programmes for both. This is an indication of how international propaganda broadcasting has become a polarised market, between East and West—and the West has more partners in its network than the East.

In the Near East, the BBC has a multi-frequency broadcasting capability from a transmitter station at Cyprus. This is equipped with two 500 kW, and one 200 kW MW transmitters. For SW broadcasting Cyprus is equipped with two 20 kW, four 100 kW and six 250 kW transmitters. Owing to its strategic location, Cyprus is one of the key relay stations of the World Service. Within range of its transmitters are large areas of the Middle East, most of Europe, and most of Africa.

In the Middle East, on the island of Masirah, leased from the Sultanate of Oman, the BBC has two 750 kW MW and four 100 kW SW transmitters. Oman is one of the last surviving colonial outposts of the British Empire, some might think. Certainly the Foreign Office exerts considerable influence there, not least by affording it some protection against subversive infiltration. Perhaps it is not surprising that the Foreign Office has a longstanding relationship with Oman as regards its use as a propaganda outlet.

In the Indian Ocean, on the Seychelles, the BBC World Service has another SW outlet: a pair of 250 kW transmitters commissioned as recently as 1988. This station provides cover for East Africa. In the Far East, at Kranji in Singapore, there is the largest concentration of all its HF relay stations, equipped with four 100 kW and five 250 kW SW frequency-agile transmitters.

In Autumn 1987, the BBC pulled a switch and brought into service another powerful SW complex, equipped with two 250 kW transmitters of the latest type, fully automated for round-the-clock broadcasting with unattended operation. As transmitting stations go, it is not the largest by any means; what is significant is its location.

Officially it is known as BBC East Asia Relay Station. In its broadcasts it merely identifies the station as BBC Eastern Relay, with no indication as to its precise location on the shore of Hong Kong island at Tsang Tsui, not more than 3 km from China. Officially the BBC says that its purpose is to project a strong signal into Japan and Korea, but another statement says that the station is punching a signal into the People's Republic of China with a clarity and strength as good as a MW signal.

There is something strange and a little sinister about this radio station, conceived, planned and brought into service at the very time when Britain ought to be thinking of shrinking its role in propaganda broadcasting in the Far East. At the very most, the British government will only get seven years' service out of this station when presumably it will be taken over by China. But a lot can happen in that time—wars have been won in much less time—and the war in the Far East is undoubtedly against the expansion of the Chinese communism.

The expansion of the BBC World Service owes much to satellite technology. Cyprus was the first relay station to be fed with programmes from London through the medium of satellite communications. In Singapore and Hong Kong, the BBC hires satellite facilities from the local telecommunications authorities. On the Ascension Islands, on Masirah, in Antigua in the Caribbean and in the Seychelles, the BBC built its own earth stations.

Using satellites in this way has represented a major advance in the audio quality of the programmes; it means that listeners on the other side of the globe can hear radio programmes that are better than many listeners to AM are able to enjoy in many parts of the British Isles. Considering how the other part of the BBC (national broadcasting) has run back its investment in national AM broadcasting, this is not so difficult a task.

Much of the strength of the BBC World Service lies in direct broadcasting by short waves in the HF spectrum. It has four high power, SW transmitting complexes: Daventry, Rampisham, Skelton and Woofferton. Daventry in the Midlands is the oldest station in the BBC's history of foreign broadcasting. Today it is equipped with four 100 kW and six 300 kW transmitters. Daventry covers the whole of Western and Eastern Europe, the USSR, the Middle East, Africa and the Americas.

Rampisham in Dorset is the most powerful of all the BBC SW stations, with two 100 kW, and ten 500 kW transmitters. This transmitting station has a similar coverage to Daventry. It is also the only BBC transmitter station fitted with four 500 kW transmitters of non-British manufacture, because when they were ordered AEG was the first company with the new generation of super transmitters. However, the last two 500 kW transmitters, fitted in 1991, were made by Marconi.

Skelton in Cumberland is nearly 50 years old, having been built during the war along with Rampisham and Woofferton. Skelton today has eleven 250 kW and six 100 kW transmitters. Its area of global coverage is similar to the other three SW stations. This is a deliberate design policy of Foreign Office planning. Each station can back up the others or even replace other stations in case of local problems such as failure of service, sabotage etc.

Woofferton is the odd one out. This is not listed as a BBC World Service station, but is maintained by the BBC for VOA. It has six 250 kW and four 300 kW transmitters. At certain times, however, it carries BBC World Service programmes.

Completing the picture in the UK are two other high-power radio stations carrying the BBC World Service: Orfordness and Droitwich. Orfordness has a

Figure 25.1 *BBC World Service station, Rampisham, UK. Aerial view of station showing matrix and transmitter hall*

covert history: it only became a BBC station in the mid 1980s, before which it was operated by the Foreign Office Diplomatic Wireless Service – a name that meant different things to different countries. Today Orfordness is fitted with two 500 kW transmitters, which normally broadcast separately on 648 and 1296 kHz – the second frequency being the first harmonic of the 648 kHz frequency. Both transmitters are of German AEG manufacture.

The site of Orfordness was chosen for its electrical ground conductivity, which is excellent, and for the all-sea propagation path that it offers to Europe, being on the coast of East Anglia. In wartime this station would play a vital role as a powerful propaganda station, because of its power and because it uses medium waves, giving a greater density of reception. There are 1000 MW receivers in use in Western Europe for every 25 SW receivers.

The programmes radiated from Orfordness are not intended to be heard in the British Isles – curtain reflectors beam the signal to Europe – but it can be heard in parts of Scotland and Southern England. None of the World Service broadcasts

Table 25.1 *Output power per station on World Service network (BBC World Service)*

Station	Transmitters and output powers	SW power, kW	MW/LW power kW
Daventry SW Station	4×100, 6×300	2200	
Rampisham SW Station	2×100, 8×500		
	2×500	5200	
Skelton SW Station	11×250, 6×100	3350	
Woofferton SW Station	6×250, 4×300	3700	
Orfordness MW	2×500		1000
Droitwich LW	1×500		500
Total for UK		14 450	1500
Cyprus MW station	2×500, 1×200		1200
Cyprus SW station	2×20, 4×100		
	2×250, 4×250	1920	
Masirah Island	2×750		1500
Masirah Island	4×100	400	
Seychelles	2×250	500	
Singapore	4×100, 5×250	1650	
Hong Kong	2×250	500	
Caribbean	2×250	500	
Ascension Isles	6×250	1500	
Exchange facilities			
Lesotho	2×100	200	Exchange facilities not included in the final total given below.
Sackville CBS, Canada	2×250	500	
Bethany, VOA	3×250 intermittent	750	
Total overseas relay stations		6950	2700
Total, overseas + UK		20 400	4200

are intended for home consumption. In this respect the BBC World Service is similar to VOA, whose broadcasts, by law, are not intended for consumption in the USA.

Droitwich is used for home and overseas broadcasting. It carries the spoken word, current-affairs programmes and serious music. Droitwich is truly an information-broadcasting station rather than propaganda.

25.1 Radio from the Foreign Office

As mentioned at the beginning of this chapter, and in the chapters on wartime broadcasting, the Foreign Office has at times controlled all aspects of programmes; but as well as this, it has controlled technical aspects to the extent of having its own transmitting stations. One other authority in Great Britain designs and constructs its own high-power transmitters: the British Post Office. From 1951 the BPO Radio Branch adopted a policy of buying its high-power transmitters from Marconi and STC in roughly equal proportions.

About that time, the Diplomatic Wireless Service of the FO expanded its commitments in high-power broadcasting and began a trawl for staff to design, construct and maintain high-power radio stations, many of whom were recruited from the BPO. The first station of the DWS was Crowborough (mentioned in previous chapters); this was followed by Cyprus, Orfordness, Masirah, Berlin and Lesotho.

This network ceased to exist in name after the BBC took over the installations and responsibility for their running. The FO also operated some stations abroad whose identities are not known (e.g. in Palestine and Somalia) and of which the Public Records Office has no files. Crowborough was far from being the ideal site from a propagation point of view, and was originally chosen for its security. It was eventually abandoned and replaced by Orfordness, which offered an all-sea path to Europe.

Table 25.2 *HMGCC transmitter network at the time of being placed under the administration of the BBC*

Sender No	Station	Transmitter power	Manufacturer
201	Cyprus	20 kW	HMGCC
202	Cyprus	20 kW	HMGCC
203	Cyprus	250 kW	Marconi
204	Cyprus	250 kW	Marconi
205	Cyprus	250 kW	Marconi
206	Cyprus	250 kW	Marconi
207	Cyprus	100 kW	Marconi
208	Cyprus	100 kW	Marconi
209	Cyprus	100 kW	Marconi
210	Cyprus	100 kW	Marconi
211	Cyprus	100 kW	Marconi
212	Cyprus	100 kW	Marconi

Table 25.2 *Continued*

213	Cyprus	100 kW	Marconi
214	Cyprus	100 kW	Marconi
215	Cyprus	500 kW (2 × 250 paralleled)	Marconi
216	Cyprus	500 kW (2 × 250 paralleled)	Marconi
A1	Crowborough	600 kW (3 × 200 kW)	RCA
A2	Crowborough	100 kW	Gates
A3	Crowborough	100 kW	Gates
A4	Crowborough	100 kW	Gates
ORF1	Orfordness	500 kW	AEG
ORF2	Orfordness	600 kW	AEG
ORF3	Orfordness	700 kW	HMGCC
MF1	Masira	750 kW	Marconi
MF2	Masira	750 kW	Marconi
HF1	Masira	100 kW	Harris
HF2	Masira	100 kW	Harris
HF3	Masira	100 kW	Harris
HF4	Masira	100 kW	Harris
VHF	Berlin		
VHF	Lesotho		

Table 25.3 *Languages broadcast by the BBC since 1938*

	Started	Ended	Restarted	Ended	Still running
Afrikaans	14.5.39	8.9.57	—	—	—
Albanian	13.11.40	20.1.67	—	—	—
Arabic	3.1.38	—	—	—	✓
Austrian (German for Austria)	29.3.43	15.9.57	—	—	—
Belgian (Flemish and French)	28.9.40	30.3.52	—	—	—
Bengali	11.10.41	—	—	—	✓
Bulgarian	7.2.40	—	—	—	✓
Burmese	2.9.40	—	—	—	✓
Cantonese	19.5.41	—	—	—	✓
Czech	8.9.39	—	—	—	✓
Danish	9.4.40	10.8.57	—	—	—
Dutch (for Europe)	11.4.40	10.8.57	—	—	—
Dutch (for Indonesia)	28.8.44	2.4.45	25.5.46	13.5.51	—
Finnish	18.3.40	—	—	—	✓
French (for Europe)	27.9.38	—	—	—	✓
French (for Canada)	2.11.42	8.5.80	—	—	—
French (for Africa)	20.6.60	—	—	—	✓
French (for S.E. Asia)	28.8.44	3.4.55	—	—	—
German	27.9.38	—	—	—	✓

Table 25.3 *Continued*

Greek	30.9.39	—	—	—	✓
Greek (for Cyprus)	16.9.40	3.6.51	—	—	—
Gujerati	1.3.42	3.9.44	—	—	—
Hausa	13.3.57	—	—	—	✓
Hebrew	30.10.49	28.10.68	—	—	—
Hindi	11.5.40	—	—	—	✓
Hokkien	1.10.42	7.2.48	—	—	—
Hungarian	5.3.39	—	—	—	✓
Icelandic	1.12.40	26.6.44	—	—	—
Indonesian	30.10.49	—	—	—	✓
Italian	27.9.38	31.12.81	—	—	—
Japanese	4.7.43	—	—	—	✓
Luxembourgish	30.11.40	30.3.52	—	—	—
Malay	2.5.41	—	—	—	✓
Maltese	10.8.40	31.12.81	—	—	—
Mandarin	19.5.41	—	—	—	✓
Marathi	1.3.42	3.9.44	31.12.44	25.12.58	—
Nepali	7.6.69	—	—	—	✓
Norwegian	9.4.40	10.8.57	—	—	—
Pashto	15.8.81	—	—	—	✓
Persian	28.12.40	—	—	—	✓
Polish	7.9.39	—	—	—	✓
Portuguese (for Brazil)	14.3.38	—	—	—	✓
Portuguese (for Europe)	4.6.39	10.8.57	28.4.63	—	✓
Romanian	15.9.39	—	—	—	✓
Russian	24.3.46	—	—	—	✓
Serbo-Croat	15.9.39	—	—	—	✓
Sinhala	10.3.42	30.3.76	—	—	—
Slovak	31.12.39	—	—	—	✓
Slovene	15.9.39	—	—	—	✓
Somali	18.7.57	—	—	—	✓
Spanish (for L/America)	14.3.38	—	—	—	✓
Spanish (for Europe)	4.6.39	31.12.81	—	—	—
Swahili	27.6.57	—	—	—	✓
Swedish	12.2.40	10.8.57	—	—	—
Swiss (German, French, Italian, English)	13.4.41	8.5.41	—	—	—
Tamil	3.5.41	—	—	—	✓
Thai	27.4.61	5.3.60	3.6.62	—	✓
Turkish	20.11.39	—	—	—	✓
Urdu	3.4.49	—	—	—	✓
Vietnamese	6.2.52	—	—	—	✓
Welsh (for Patagonia)	1945	1946	—	—	—

25.2 Future expansion

For all its massive capability, with a total output power of over 24 MW on all AM wavebands, the BBC is already planning further expansion during the next decade. It is actively seeking to position a high power medium wave transmitter on Gibraltar with which it hopes to project a powerful signal into Maghreb (Arabic North-West Africa).

It is also seeking to strengthen its SW transmissions from Masirah by replacing the ageing four 100 kW transmitters with an equal number of 300–500 kW transmitters. On the same station it would like to increase its MW broadcasting capability with the addition of another 750–1000 kW transmitter to serve India and Pakistan. Another area of the world where the BBC thinks it is losing out to competitors is in South East Asia, where VOA has the benefit of a 1 MW transmitter blasting its way into territories the BBC cannot reach. The BBC eyes enviously the plans made by USIA to invest $1.2 billion in the VOA Improved Audibility Program, which originally called for the purchase of 55 super transmitters.

The financial cost to the British taxpayer for the upkeep of the BBC World Service was some £119 million in 1988–89, and is expected to rise sharply in the next few years. The BBC prefer to describe this figure differently: 2 pence per week for each of its 120 million listeners. Either way, the BBC and the Foreign Office think it a small price to pay for the inestimable credit to be gained from projecting the better side of Britain to the rest of the world.

Chapter 26

Subversion, propaganda broadcasting and the CIA

During the war, the US Office of Strategic Services (OSS) was involved in many of the murkier aspects of war, including covert broadcasting. After the war, the OSS disappeared and was replaced by the Central Intelligence Agency. Originally, as its name suggests, the CIA was a fact-finding agency, but an extraordinary law was passed in 1949 — The Central Intelligence Agency Act — which empowered the CIA to undertake unspecified activities abroad, and exempted it from publication or disclosure of the organisation, functions, names, official titles and salaries, or numbers of personnel employed by the agency. It was even allowed to spend money without regard to the provisions of the law and regulations relating to the expenditure of government funds; its budget was buried in the budgets of other agencies, and its personnel could infiltrate other agencies and organisations, government or private. The CIA became truly a product, and an instrument, of the Cold War.

The creation of the CIA is paralleled in recent history only by that of the SS by Hitler; it was a state within a state, answerable only to the Secretary of State — and the Secretary of State was the brother of the head of the CIA. John Foster Dulles and Allen Welsh Dulles were of different temperament, but collectively formed a synergistic force. Allen, the younger brother, was described as genial and socially adaptable. John Foster Dulles was earnest and self-absorbed, a lawyer by profession. When President Eisenhower appointed him to the position of Secretary of State, it was a move destined to have the most far reaching effects upon America's foreign policy.

John Foster Dulles was possessed of a hatred of communism that almost amounted to an evangelical mission. The Truman policy of merely containing communism was swept aside for a much stronger one of 'Liberation'. The iron curtain would be rolled back, the people of the captive nations no longer abandoned to godless terrorism, 'unrelenting pressures would make the communist rulers impotent to continue in their monstrous ways.'

Yet despite this rhetoric, John Foster Dulles did not apparently show much interest in the activities of the Voice of America. He had in mind an altogether different kind of broadcasting agency. One of his first appointments was to appoint General Bedell Smith (then head of the CIA) to the position of Under Secretary of State. This created a vacancy for a new top man for the CIA to which his brother

Allen was appointed. Allen Dulles was well experienced for this role: he had held the top position in the OSS during the war. As a result of these staff changes there was formed a strong link between the CIA and the State Department itself, with the Dulles brothers controlling both. John Foster Dulles, as Secretary of State, also had complete charge of the USIA—which in turn controlled the VOA. He now had the power to control every aspect of international propaganda broadcasting, and this he did by setting up a rival propaganda broadcasting agency: Radio Free Europe.

The idea of having two broadcasting agencies—one of overt character, the other covert—was not original; it was a continuation of the wartime policies exercised by Britain and the USA. Although this was supposed to be a time of peace, any weapon that might help in destroying communism was justifiable to the US Government. As well as VOA and RFE, there were the national, privately owned broadcasting companies such as CBS, NBC and ABC, on which the government could call—and did—when it felt such voices were needed.

RFE was the first of many such covert propaganda broadcasting agencies. The broadcasting empire over which the Secretary of State presided was global in size and complex in its politics. The State Department exercised indirect control, through the FCC, of the national broadcasting networks such as CBS, NBC, ABC and others, each of which operated as many as 12 radio stations. Coexisting with these national broadcasters was the Voice of America, an overt and respected international broadcasting agency under the control of the USIA. Also coming under the direction of the State Department was the US forces broadcasting service AFRTS.

The role of the CIA's broadcasting service was altogether different; international broadcasting networks such as Radio Free Europe and Radio Liberty were covert broadcasters operating under the guise of privately funded radio stations. RFE, when first formed in the early 1950s, was announced as a private station financed by donors. This was in fact only a cover; a small fraction of the costs came from private sources, the rest from the CIA. The setting up of RFE was highly successful. Its headquarters was in Munich and its staff consisted of anti-communist intellectuals recruited locally in Eastern Europe from Poland, Czechoslovakia, Hungary, Bulgaria and Romania. These dissidents were welded into a cohesive, powerful force for the purpose of levelling in-depth critical analysis at the communist regimes. The style of programmes was in marked contrast to that of the VOA: RFE was less restrained in its aggression.

However, in the early 1950s, VOA itself was more strident than today. In 1953 it made its bid for world stardom when it brought into service a giant 1 MW LW transmitter in southern Germany with the avowed objective of rolling back the frontiers of communism. This action by the US Government was an act of international piracy on a giant scale: it used the frequency of 173 kHz which was that officially allocated to Radio Moscow, it was an unlicenced broadcasting station operating in another country, and it used power on a larger scale than any other radio station in the world.

By 1955 RFE had acquired some 29 powerful radio transmitters in places including Portugal, Spain and West Germany. RFE became the model for other covert broadcasting networks, such as Radio Liberty and Radio Free Asia. America, through its many broadcasting agencies, had put into effect a 'ring plan', with a string of powerful radio stations encircling the earth, but always situated as near as possible to Soviet soil: there was no place in the Soviet empire that US propaganda stations could not reach with ease.

An even darker side to the CIA's fight against communism emerged in the Middle East, where the Arabic-speaking countries had expressed a desire to use radio broadcasting as a means of spreading the voice of Islam to their peoples. The constructing of powerful radio stations, transmitters, antennas and generating plants is an expensive undertaking for any country, but particularly those who must pay the costs of importing this high technology. Then there is the cost of running these super transmitters. A modern, high power radio station might need up to 40 000 gallons of oil per day, to generate the required electrical power to run the station. In some instances the high capital costs and the running costs were beyond the means of some Middle East countries.

Such countries usually sought funding through either Soviet or US aid programmes. It was not unknown for a secret agreement to be reached for the funding nation to have access to the radio transmitters at certain times of the day. As recently as 1980, the CIA arranged with the Egyptian government that a super transmitter, supplied and installed with US aid, should carry an anti-Khomeni service known as 'the Free Voice of Iran'. The object of these transmissions in 1980 was to aim for the overthrow of the revolutionary Khomeni government in Iran.

This was only one of many instances where the CIA became involved with subversive propaganda broadcasting. The most notable was probably 'Radio Swan'. This was a 50 kW CIA station constructed on Swan Island in the Caribbean. This was to be the 'liberation' station to foment rebellion in Cuba. It went on the air in May 1960—the same time that the USA and USSR agreed to defuse the Cold War. The invasion of Cuba by American forces was a fiasco, but Radio Swan lived on to be renamed as Radio Americas.

In the twilight world of subversive propaganda broadcasting, it is a fairly reliable assumption that any radio station with the words 'free', 'peace' or 'liberty' in its name is designed to create unrest or bring about a revolution. It has now been established that the Hungarian uprising was brought about by Radio Free Europe. On 27 October 1956, RFE broadcast a programme by a Colonel Bell, which carried the message that military assistance would follow an uprising. On 30 October another Hungarian speaker gave instructions in anti-tank warfare, quoting experiences from the Second World War.

By 4 November the uprising had reached a stage where European newspapers were predicting a Soviet invasion. On that day, a speaker on RFE said: 'In the Western capitals, a practical manifestation of Western sympathy is expected at any hour.' This was the clearest indication that the revolt should go ahead. After this, many Hungarian amateur radio users sent messages asking for assistance. One typical from one of these stations broadcast the last message: 'Attention, Radio Free Europe. Attention! We request immediate information. Is help coming from the West?'

Sadly, the West had no intention of coming to the aid of those it had encouraged to revolt. Radio Free Europe had achieved its objective: the destabilisation of an East European nation. Shortly after this episode the CIA changed the name of the sister broadcasting agency Radio Liberation to Radio Liberty—possibly to avoid any indication that its purpose was to liberate.

The Hungarian uprising did nothing to improve relations between West

Germany—where RFE transmitters were located—and the staff of RFE, who had become embroiled in the bloodbath of Hungary. RFE was later to publish the result of a survey, which showed that of 1107 Hungarian refugees who were asked whether they had expected Western support, 96% said yes.

Since the depths of the Cold War in the 1960s when a nuclear confrontation between the USA and USSR came close to reality, the tone and rhetoric of Radio Free Europe and Radio Liberty has lost a little of its harshness, although it still projects a more belligerent voice than the more refined VOA.

Notwithstanding an easing of tension between East and West it might be premature to believe that it will always be so. President Nixon in his 'The real war' (1980) makes it clear that he regards world war III is already in progress, and that the US is engaged in this with the Soviet Union.

p. 341 'We regard the Soviet frontiers as the frontiers of our own defences.'

p. 313 'There are many things that detente cannot do, it cannot turn the Russians into good guys.'

p. 251 'One thing we should not expect is that the Soviet Union is going to mellow, or that the Soviet values and ambitions are going to be markedly different from what they have been'

p. 41 On terrorism: 'The Soviet Union's inhuman contempt for even the most basic of civilised standards'

p. 286 On the need for secrecy and blackout: 'The black arts should be restored to dimmer light, or in darkness where they cannot be seen'

p. 286 'Unless we restore the covert capabilities of CIA we are going to get rolled by the Soviets...', 'The Freedom of Information Act lets anyone rummage through CIA files'

p. 338 'We should not shrink from the propaganda war, either within the Soviet empire or in the rest of the world. We should revitalise Radio Free Europe and Radio Liberty and set up counterparts of them that can compete directly with Soviet propaganda.'

26.1 RFE/RL: expansion

According to the most recent published data, Radio Free Europe/Radio Liberty Inc. is a privately owned company, address Oettingenstrasse 67, Munich 22, whose president is named Gene Pell. It has offices in Washington and New York. Its radio transmitters are located in Germany, Spain and Portugal. According to other published information, RFE/RL is a non-profit-making, privately owned and managed radio network broadcasting news and information to the peoples of Bulgaria, Czechoslovakia, Hungary, Poland, Romania and the USSR.

It would be pushing human credibility to suppose that any person or group of persons acting philanthropically would go to the expense of setting up super transmitters for the purpose of international propaganda broadcasting. Possibly because of this, the US government makes little secret of the fact that RFE/RL is funded by US government appropriations through the Board for International Broadcasting.

Presently, RFE/RL has three transmitter sites in Germany (Holzkirchen, Biblis, and Lampertheim), plus Playa de Pais in Spain, and Gloria near Lisbon. The transmitters and power ratings are summarised in Table 26.1.

Table 26.1 *RFE/RL transmitters and powers*

Site	Transmitters
Holzkirchen	150 kW MW
Biblis	9 × 100 kW SW
Holzkirchen	4 × 250 kW SW
Lampertheim	7 × 100 kW SW
	1 × 20 kW SW
Playa de Pais	5 × 250 kW SW
	1 × 100 kW SW
Gloria	19 × 250 kW SW
	2 × 100 kW SW

With a total transmitter power of 8970 kW, RFE/RL is much smaller than the VOA broadcasting network. However, a simple comparison such as this could be grossly misleading unless the role of the broadcasting agency is taken into account. VOA is a global broadcasting network with a global role; while its principal target is Russia, it needs to be born in mind that the landmass of Russia stretches around the world from Europe to the Pacific.

RFE/RL was set up to carry the voice of American propaganda solely to Eastern Europe. RFE transmissions are broadcast in Bulgarian, Czech, Slovak, Hungarian, Polish, Romanian, Latvian, Lithuanian and Estonian. Radio Liberty transmissions are broadcast to certain ethnic groups in Russian, Ukrainian, Armenian, Azeri, Byelorussian, Georgian, Tatar, Bashkir, Kazakh, Kirghiz, Tadjik, Turkoman and Uzbek. A transmission is also directed towards Afghanistan.

The creed of RFE/RL is a slogan adopted from the universal declaration of human rights: 'Everyone has the right to seek, to receive, and impart information regardless of frontiers'. On the other hand, an earlier 1936 Geneva resolution condemned radio broadcasts designed to incite the population of any territory to acts incompatible with internal order: much of the propaganda broadcasting carried on around the world is with this objective.

Until the USSR suspended its radio jamming operations in May 1989, in line with its new policy of *glasnost*, the RFE/RL transmissions had the distinction of being the most heavily jammed of all transmissions in the HF broadcasting bands. According to VOA sources, no fewer than 15 different language broadcasts had been jammed by the Soviet authorities, using a combination of ground-wave and sky-wave jamming stations in the USSR and various Eastern Bloc countries. From the same source of published information, RFE/RL emerges as the world's most heavily jammed broadcaster, followed by VOA in second place. Other international broadcasting agencies whose broadcasts have been jammed by the Soviets are those of China, Germany, Greece, Israel, Korea, Italy and the Vatican City.

In May 1989, the Soviet news agency Tass reported: ' "The decision to stop jamming broadcasts is another step forward in the development of *glasnost*," said the Minister of Communications of Byelorussia Ivan Gritsuk. "New conditions in international broadcasting must prompt all participants. . . to greater responsibilities for their activity, based on the principles of humanism and co-operation, and of discarding the stereotypes of the cold war." '

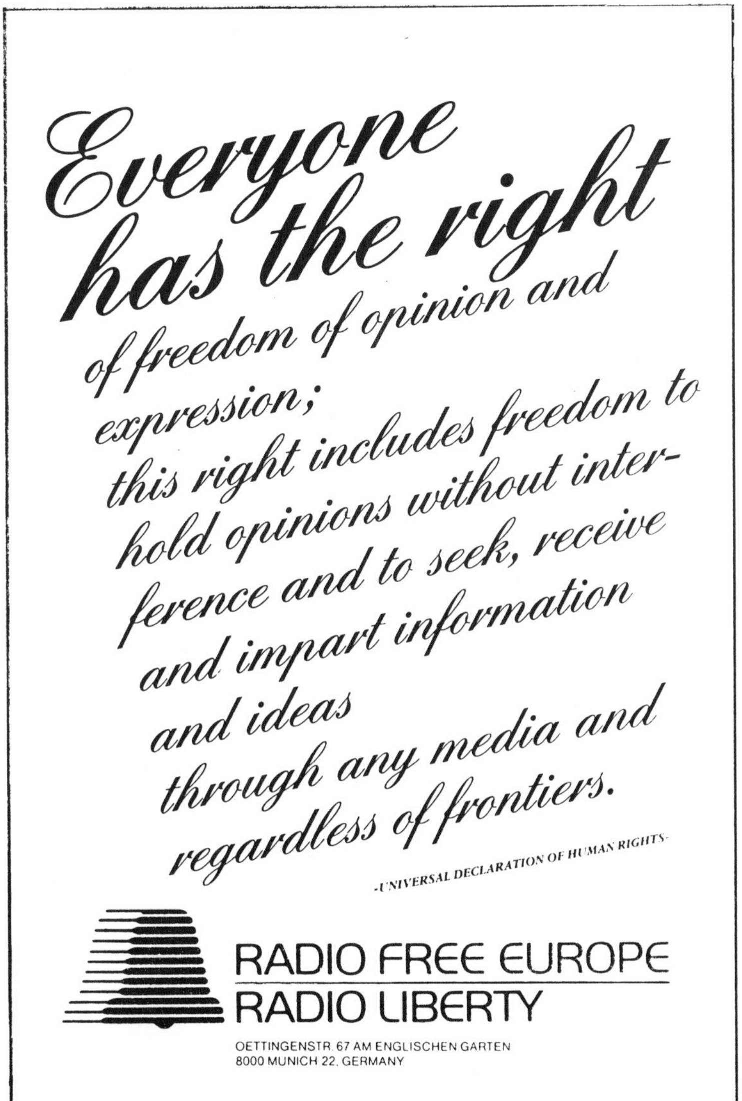

Figure 26.1 *Advertisement in WRTH for Radio Free Europe/Radio Liberty*

Given the continuation of propaganda broadcasting, and the flow of rhetoric of the kind mentioned earlier, it may be premature to believe that the suspension of jamming even now is permanent. The Cold War has been characterised by many phases of a non-permanent nature. The Soviet suspension of jamming operations did not bring about a reciprocal gesture from the USA; quite the reverse. In September 1989, Asea Brown Boveri announced the award of a contract from RFE/RL Munich headquarters for four 500 kW high performance super transmitters for the services' Portugal site.

Earlier in 1989, Thomson-CSF announced from its headquarters in Paris that it had been awarded a contract from the US Board of International Broadcasting for the supply of three 100 kW SW transmitters, to be installed in Biblis and Lampertheim transmitter sites. RFE/RL are also making changes and design improvements to the SW antenna arrays at these sites to improve the overall performance of the transmission network.

26.2 The Dulles family

During the 1950s, the Dulles brothers together with their sister effectively controlled the European policy of the USA. John Foster Dulles was Secretary of State, Allen Welsh Dulles was Director of the CIA and Eleanor Lansing Dulles was in charge of the Berlin desk at the State Department. All three assumed that they would shape the role of America in its battle against communism.

Allen Dulles, by virtue of his position, was, the Soviets thought, the most powerful and the most dangerous man on earth. He was all-powerful with an unlimited budget; he ran his own army of spies, his own mercenary army, his own planes, he recruited his own killers, he organised nearly a hundred revolutions in other lands. He also ran his own broadcasting empire, dedicated to the overthrow of communism using methods and techniques that have no place in the armoury of a civilised society.

The Dulles dynasty lasted for ten years. When John Foster Dulles died in 1959, his role as Secretary of State was taken over by a new man: the more moderate Christian Herter. There was then nobody to protect Allen Dulles and the CIA from criticism after the Bay of Pigs disaster. By the time John F. Kennedy took office, the writing was on the wall: the man who had masterminded a hundred major coups, who had humiliated and outwitted Khruschev, manipulated foreign governments and toppled dictators, was now about to be toppled himself. He was eventually dismissed in 1961.

Chapter 27

Second in the world: the USSR

Russia has a claim to antiquity in the history of broadcasting that is probably unequalled by any other nation. In Russia there is no doubt that it was a Russian, Professor Popov, who first invented radio, in 1895—one year before Gulgielmo Marconi came on the scene. The Russian naval fleet was equipped with wireless during the First World War, and the first ever public broadcast took place in October 1917 from the Russian cruiser 'Aurora' moored on the Neva river within sight of the Winter Palace in St Petersburg.

Home service broadcasting began in 1922—the same year as in many other European countries. Foreign service broadcasting started in 1929, with broadcasts in German, English and French; the Soviets were quick to recognise the importance of broadcasting in several languages, a fact that did not occur to the BBC for another decade, as it persisted right up to 1938 in broadcasting to the world in English only.

By 1941 the USSR was broadcasting to the world in 21 languages; at its peak, it broadcast in 64 different languages, and measured by the number of hours per week it grew to the second largest international broadcaster in the world, with 2257 hours of radio programming, compared to the USA's 2368 hours.

It is important to remember the geography of the old Soviet Union. It extended from Western Europe—the Baltic States—to the easternmost tip of Asia. When it is 5 a.m. on Cape Dezhnev in the Far East, it is midnight on Lake Baikal in Siberia, while in Moscow it is only 7 p.m. on the previous day. It stretched 172° of longitude from east to west—nearly half-way round the globe. With no fewer than 11 time-zones, the Soviet Union was immense.

Russia alone is still larger than any other country; it could accommodate quite easily the whole of the USA within its area and its population is roughly 137 million, nearly half of that of the USSR as a whole. The products of a country tend to reflect the character of that country; in the same way the Texans build large automobiles, the USSR excels when it comes to heavy engineering and electrical engineering. It outperforms the rest of the world in its size of ships, icebreakers and helicopters, whilst at the Soviet Research Centre at Novosibirsk, Soviet scientists are perfecting the 3 MV super-tension grid system. Developments in miniaturisation and micro-miniaturisation are things that the Russians would prefer to leave to the Japanese.

The number of countries that carry out propaganda or information broadcasting directed to other countries grows with each passing year, and is currently in excess of 100. Of these about 30 are the principal broadcasters. In this league the top eight performers are:

Table 27.1 *Major broadcasters of the world*

USA	2368	42 languages
USSR	2257	64 languages
People's Republic of China	1517	
German Federal Republic	831	
United Kingdom	756	38 languages
Egypt	549	
North Korea	548	
East Germany	480	

Whilst the figures for the two super-powers seem to be evenly matched in terms of programme-hours per week, this is not the only parameter to be considered; it is merely a yardstick which takes no account of the technical factors. These are a little more complicated, and relate to numbers of transmitting stations, numbers of transmitters at these stations, output power rating of the transmitters, and finally their strategic location with reference to the target zones of the transmissions.

To be effective, propaganda broadcasting requires a strong signal to arrive at the target zone. The attenuation characteristic of radio signals is a function of the distance from the transmitter to the receiver. It is also subject to the vagaries of the ionosphere, and the number of hops that the radio signal traverses. SW broadcasting is at its most reliable when the distance from the transmitter to the receiver location is less than 2000 km. Western powers, such as the USA and UK, had a distinct advantage in this respect: their worldwide connections enabled them to negotiate agreements with countries around the world for siting propaganda radio stations in strategic locations close to the target areas.

Today, with a VOA broadcast, the listener may be unaware that the transmission is coming from anywhere but the USA—it might be from Germany, Thailand, Turkey, Greece or wherever. By contrast, with Radio Moscow the listener may be assured that the transmission is from Russian territory; this means that the signal may have travelled several thousand kilometres. Yet the fact remains that the voice of Radio Moscow can be heard reliably in all parts of the world, provided that the listener has selected the best frequency. Radio Moscow achieves this by the use of powerful transmitting stations.

Powerful radio transmitters have always been a factor of Soviet foreign broadcasting; even in the early 1930s, Radio Moscow could be received with clarity and strength all over Europe.

All broadcasting in the USSR was state controlled, and like most countries it had one authority responsible for all home broadcasting, and another authority for all external broadcasting. The body responsible for all radio and television broadcasting within the USSR was the State Committee for Television & Radio Broadcasting, based at Ulitsa Akademika Koroleva, Moscow. External broadcasting was controlled by the Ministry of Posts & Telecommunications, Department of External Relations, Moscow. Each republic also had its own minister.

Each of the 15 republics had its own broadcasting network, in keeping with the size of the different republics. There are still super-power stations, high-power stations, medium and low-powered radio stations—too numerous to list here, and indeed it is difficult to find anyone in the USSR who could identify all the stations.

The numbers of different frequencies allocated for AM broadcasting on the long and medium wavebands alone for national broadcasting total over 120 frequencies as at 1987. Available figures for 1987 show over 21 stations with carrier powers of 1000 kW or more, 27 with powers of 500 kW, and many more with 250 kW transmitters. External broadcasting is sometimes carried out over national and regional transmitting stations, most commonly to the Middle East and North Africa.

27.1 Transmission technology

Soviet transmission technology is something of an enigma. For nearly three years, I wrote to the Soviet government requesting permission to visit some high-power SW transmitting stations, without receiving a single reply. After these futile attempts I contacted Tass news agency in London, who told me: 'Soviet technology is a paradox—in some things they are still in the Victorian age, in other things they are leading the world. In the mind of the Kremlin both are very good reasons for refusing to divulge any information.'

Although the situation has since changed, and information is easier to obtain, it is still extremely difficult to obtain specific answers to specific questions. Short of examining every high-power transmitter in Russia and her erstwhile satellites—impossible task—one can only generalise on what information has been found out.

All transmitters were built by the state industries, and do not appear to be based on current design techniques found in the West. The paradox here is that, although the USSR invests more in fundamental research than the USA, the results of this research seem to be channelled into a fairly narrow field of technology, and radio broadcasting is not one of these priorities. A 2000 kW MW transmitter will generate the same field strength whether it is modern or ancient, all other things being equal.

All international broadcasting agencies attach great importance to continuity of transmission service; this is one of the factors that led to the modern frequency-agile transmitter, with the ability to be re-tuned to a new frequency in the shortest possible time. This operational requirement led Western manufacturers towards the auto-tuning transmitter. But the same effect can be achieved by having many more transmitters of the manually tuned variety, each tuned to a different frequency. In the West, such methods were abandoned over 25 years ago because of the space requirements, and of the numbers of transmitters that were needed. In the USSR restrictions on space and numbers of transmitters do not apply.

It was rising energy costs, after the oil crises of the 1970s, that spurred the West into producing more efficient designs of radio transmitters with high overall conversion efficiencies, best exemplified by the development of pulse-duration modulation and pulse-step modulation, which produced an even higher overall conversion efficiency. Whether because the USSR had its own oil reserves or some other reason, the Soviet Union produced nothing to match the technology that was developed by Western manufacturers such as AEG, ABB, and Thomson.

Thus the Soviet-built high power transmitters feature technology such as class B modulators, and the liberal use of valves or tubes. This makes for bigger and more cumbersome units. A Soviet-built 500 kW transmitter occupies over twice the floor area of a modern European designed transmitter. Similarly, Western broadcasting agencies make use of SW antennas that are space-conserving whilst still having high-gain characteristics. The four-by-six HF curtain arrays are a good example of this. By comparison, Russian stations use considerable numbers of rhombic antennas. Whilst this type of antenna is capable of a high gain of 20 dB, it is a design that takes a lot of space: a fact that is not so much of a disadvantage in Russia.

The technique of combining outputs from more than one transmitter with the object of increasing the field strength at distant points is much in use in the USSR. To retain a high quality it is necessary to phase the outputs of the transmitters, but it is possible to dispense with this refinement. The Russian broadcasting systems, both national and international, take advantage of every technique to increase the radiated carrier power. It is quite common for as many as 20 transmitters located at different sites to broadcast the same programme on the same frequency, giving a total amount of power as high as several megawatts.

Unlike VOA, which has transmitters located at strategic points worldwide, the USSR never spread its stations around the world. One could argue that given its size, the Soviet Union had no need of further strategic bases elsewhere in the world. Apart from the countries of Eastern Europe, the USSR broadcast through only one other outlet: Radio Habana, Cuba. In recent years the USSR suspended its agreements with Fidel Castro to relay Soviet programmes over that station, as part of its efforts to improve Soviet-American relations.

By means of its size, the USSR was able to compete with VOA in most areas of the Middle East, North Africa, the Near East, South Asia, South East Asia, China and Australasia. The region of the world where it could not reach was the Americas. The USA itself is a country that has almost abandoned the practice of SW listening, which made it difficult for the USSR to project its culture through SW broadcasting to the USA.

Chapter 28

Renewed expansion at the Voice of America

VOA was born in 1942, in the dark days of World War Two. As America's voice to the world, its mission is to report the news to listeners whether that news be good or bad. It has never wavered in the belief that its broadcasts carry the voice of truth. This same conviction is also held by the other major broadcasters in the world: the BBC World Service, Deutsche Welle, Radio Cairo, Cuba Radio, the People's Republic of China, and many others.

Truth is an elusive commodity, perhaps best expressed by Oscar Wilde as being 'Never pure and rarely simple'. As one man's freedom fighter is another man's terrorist, so information broadcasting becomes labelled as truth or lies according to the source of the broadcast. Sometimes the truth turns out to be lies and that which has been labelled lies all too often is later revealed as the truth.

The importance attached to information broadcasting has never been higher; world expenditure is increasing rapidly as more emergent nations become aware of the potential. Broadcasting in the HF spectrum is a nation's fifth estate, enabling it to extend its zone of political influence into territorial areas of other nations – a feat that in the 19th Century would have been possible only with an invading army. Some nations are reluctant to admit that international broadcasting carries a higher degree of priority than its national broadcasting. This is not the case with the USA. The Voice of America is unique in other ways: US domestic broadcasting is carried out by private enterprise, but international broadcasting is controlled by the US Information Agency (USIA). Secondly, its broadcasting policy goes hand in hand with the defence policy of stationing forces in host countries. There are many advantages to be gained from this policy; apart from reduced running costs it extends the political boundaries of America by several thousand miles.

Thirdly, the USA makes no secret about its objective, which is to project a clear and powerful signal to all parts of the world. In July 1982, in a typically overt statement, President Reagan announced:

> 'We intend to move forward, consistent with budgetary requirements, to modernise our primary means of communication – our international radio system. The Voice of America, Radio Free Europe and Radio

> Liberty have been neglected for many years. Their equipment is old and deteriorating, their resources strained, and little has been done to counter the jamming that has intensified in recent years.'

Again in 1984, at the signing of an agreement with Morocco on the extension of VOA's broadcasting capability from that country, President Reagan declared:

> 'The Voice of America has been a strong voice for truth. Despite problems of antiquated equipment and Soviet jamming, the Voice of America has been able to extend its message of truth around the world. Were it not for many years of neglect, the Voice of America could be heard more clearly by many more people around the globe. And that's why our administration has made the same kind of commitment to modernising the Voice of America that President Kennedy brought to the space programme.'

The neglect that President Reagan was speaking of was of a technical nature rather than geopolitical. This network of interlinking satellites, microwave, SW and high-power MW broadcasting stations encircles the earth. About the only things of true US origin are the programmes from the VOA headquarters in Washington, for the broadcasts themselves are almost certainly emanating from locations thousands of miles from America.

The Americans have always exhibited great talent in the planning of global strategic networks, whether military, civil aviation or communications. VOA extends over every continent and embraces installations in 15 host countries, including the UK. Yet America forbids the siting of foreign radio stations in its territories.

It is virtually impossible to overstate the advantages of such a worldwide integrated broadcasting network. SW broadcasting is the most effective way of communicating over long distances, but it is still plagued by difficulties brought about by multiple-hop reflection. It works best at a range of less than 1600 km, and at greater distances where the intermediate reflection point is over sea. Conversely, when it reflects over desert regions, the signal may be further attenuated by as much as 15 dB.

Another transmission medium coming into play for international broadcasting is the MW band. Frequencies within this waveband have a different characteristic to the short waves, but at night they can exhibit a similar effect to the short waves in that they are subject to ionospheric reflection, extending the serviceable range. Night-time broadcasting at the upper frequencies of the MW band comes into its own for across-the-border, medium-range broadcasting, ideal for densely populated areas such as Europe.

Whilst VOA aims to project American culture to every part of the world, it makes no secret of the fact that its main target has always been Eastern Europe. For the foreseeable future, SW broadcasting will continue to be the mainstay of its operations, supplemented by a smaller number of high power medium wave transmitters at strategic locations.

To encircle the USSR, VOA has five main bases in host countries. Woofferton in England, ostensibly a BBC transmitter station, is funded by the USIA. With a total of 2700 kW of SW power and 37 high-gain directive antennas, this station serves north-eastern Europe.

Further south is the VOA station in Tangier with 440 kW, and Munich with an output of 540 kW. Both of these stations are due for an upgrading in transmitter output power. In the Near East VOA has a SW station at Kavala in Greece, with a massive total of 2750 kW and 23 directive antennas, all directed towards the former Eastern Bloc countries. However, the most powerful overseas SW station of the VOA network is located at Tinang in the Philippines.

The most striking features of this worldwide transmitter network are the variety of different manufacturers' equipments and the age of some of these transmitters. Even after allowing for the fact that the economic life of any high-power transmitter is measured in decades, over 35% of its network is over 30 years old—some transmitters are of wartime vintage. Network technical deficiencies extend to more than just age. VOA has lagged in the state-of-the-art in such things as frequency agility, super power and new modulation techniques to permit high efficiency operation.

This was what Ronald Reagan was referring to in his speeches. Since then the USIA has procured four transmitters incorporating all of these features, each from a different manufacturer, for the purpose of a 12 month technical evaluation. Previously, VOA had no 500 kW transmitters, compared with eight in service with the BBC, 11 with Radio France International, nine in service with Deutsche Welle and 33 in the USSR. VOA before 1985 might be compared with a slumbering giant: its geographical infrastructure is huge, but its technical muscle is weak. All this is going to change in the future when VOA acquires its new transmitters.

The USA has never been a leader in original research, and this applies to its involvement with newer technologies in high power transmitters. With its policy of unregulated domestic broadcasting, the USA has more equipment manufacturers than any other country in the world, but since, as a rule, owning a domestic radio station in the USA is very much like owning a petrol station—in fact many have made the switch—most privately owned radio stations tend to be of the lower-power variety. The two leading companies are Harris and Continental Electronics of Dallas.

Harris is a high technology company, but has so far restricted its AM transmitter products to a power of 100 kW. Continental, on the other hand, has built some of the biggest transmitters in the world for MW use, but has only recently ventured into the more sophisticated field of the frequency agile, automatically tuned transmitter with carrier powers of 500 kW.

The design and development of such equipments is an activity that calls for much experience and heavy investment in research and development, and carries no promise of massive profits at the end. Because of this, it is the international giants such as AEG, GEC and ABB that are the front runners.

In 1984, the USIA set up an Office of Co-ordination for international negotiations. It was the task of this department to work with foreign governments on negotiations for bases and to co-operate with VOA's radio engineering department, to identify those areas of the world where the 'Voice' is not sufficiently strong. Since the inception of this new department, successful agreements have been concluded with the governments of Sri Lanka, Thailand, Costa Rica and Belize. More are to follow but the process of negotiating transmitter sites in a host country is a delicate one—reciprocal agreements for trade, economic aid, and other 'fringe' benefits can often bring about a conclusion that is satisfactory to both countries. It is usually the less wealthy developing nations which the USA seeks out for its VOA locations—the exceptions, such as the UK and Germany, honour long-standing agreements dating back to the war and immediately after.

28.1 The great expansion

In 1985, VOA was experiencing interference from Soviet jamming operations, and at that time no one foresaw its ending. In such a cold climate, it is easy to understand the massive expansion programme that VOA set itself. VOA issued a tender document for 100 super transmitters. The 'Request for Proposal' document consisted of 30 heavy volumes of tender documents and performance specifications. The companies invited to tender had never seen anything so massive. The sheer numbers of high power transmitters that were called for were beyond the capacity of all the manufacturers to produce in the required time, when other existing orders were taken into account.

Quite clearly, VOA was intended by the State Department and the USIA to dominate the airwaves and drown the voice of its enemy. Since there seemed to be little prospect of any single company being able to fulfil the order, it seemed to be a natural assumption to think the total requirement would be split between at least three companies.

In 1987 after the VOA technical team had evaluated models from AEG, Brown Boveri, Continental Electronics and Marconi, a contract award seemed to be imminent, and the signs were that the major manufacturers of Europe who had put considerable effort in the answering of all technical matters would now share in a contract award.

This impression was given further reinforcement when I received the following letter, in the spring of 1987, from the VOA in response to an enquiry on progress with the 'Increased audibility programme':

> 'We purchased four 500 kW transmitters for evaluation, these were supplied by Marconi, Continental, Brown Boveri and AEG/Telefunken. All of them were able to meet their specification and are now in regular service. We are pleased with their ease of operation, efficiency, and low failure rate. We are therefore looking forward to the installation of similar transmitters in our new locations, and eventually in our old stations.'

There the matter of the contract award rested. For another year nothing happened, and in the meantime the various contractors nervously refrained from giving promises of delivery of similar transmitters to other broadcasting authorities, in case they suddenly received the big order from the USIA. When eventually the USIA made its long-awaited pronouncement, it took everyone by surprise. During the long wait there was mounting speculation on the result. To most, there were two favourites: the Swiss company Brown Boveri and the American contender Continental Electronics. Brown Boveri had already supplied VOA with nine 250 kW SW transmitters of advanced design, which were performing satisfactory service in the Philippines and at other locations. Brown Boveri also possessed the most efficient design—the only one with pulse-step modulation. The case for Continental was that it had supplied more of its high-power transmitters to

VOA than any other company – 32 in all – and 21 of these had carrier powers ranging from 250 to 1000 kW. It also held the record for supplying the highest power in transmitters: two 1000 kW MW transmitters in the Philippines and Thailand.

An examination of the transmitter records for VOA overseas stations shows that over a period of 40 or so years it had procured transmitters from no fewer than 15 different suppliers: practically every major manufacturer in the USA and a few from Europe. It looked as if VOA was seeking the perfect supplier, and that Continental was the front runner. But on 31 May 1988, the USIA announced that the contract had been awarded to GEC-Marconi – the one company that had not supplied any transmitters to VOA, apart from the single evaluation model of its B-6127.

The contract award included options for a total of thirty 500 kW SW transmitters and two 500 kW MW transmitters – the highest valued contract in the history of broadcasting. Instead of splitting the contract between several companies, the USIA had awarded the whole contract to GEC-Marconi. GEC had set up in the USA a consortium of two companies: Marconi Electronics Inc. and the Cincinnati Electronics Corporation, which would manufacture the Marconi-designed 500 kW transmitter in the USA.

From the viewpoint of the US State Department and the USIA it had secured the best deal. At a press conference held in Washington after the award was announced, it emerged that the transmitters were to be built in the USA from the Marconi B-6128 design developed at Chelmsford, England. The RF subsystems would also be built in the USA, altogether totalling over 80% of the contract value. In effect, then, it was a 'technology transfer' arrangement, benefiting the British little except for prestige.

The rest of the bidders greeted the news with some consternation. Reverberations continued among the transmitter industries for months. All had spent a great deal of company effort over some years in protracted discussions with VOA, and co-operated in supplying evaluation models of their transmitters, which were installed at VOA headquarters station, Greenville, North Carolina – all of which had met the specifications.

The 12 month evaluation programme was conducted by a Technical Evaluation Board and a Cost Evaluation Board, comprising experts from various fields. The recommendations of these boards were then reviewed by a Source Selection Advisory Council. All this emerged at the press conference referred to earlier. It was also revealed that technical aspects were evaluated with price, and that 'price carried twice the weight of technical merit'.

Brown Boveri had produced an outstanding design of transmitter whose performance had exceeded the efficiency requirements laid down in the RFP Document Tender. According to the VOA Evaluation report an overall efficiency of 85% average was achieved against an average of 65% for the other manufacturers' equipments. On balance, however, it was probably the Dallas-based company Continental Electronics that suffered the biggest disappointment – and a little humiliation, for it had served successive US administrations for almost 30 years in supplying high-power transmitters to VOA and Radio Free Europe since the early years of the Cold War. It is a speculative thought that if the Bush Administration had been in power things might have been different on the final outcome. California and Texas are two great rival states that exercise considerable

Figure 28.1 *Continental 420B 500 kW shortwave transmitter installed at VOA, Greenville, USA*

influence; Reagan is a Californian and Continental headquarters is at Dallas, Texas.

International propaganda broadcasting is about world politics and politics does play a part in contracts. It is more than likely that the 'special relationship' between the UK and USA played its part in the VOA contract.

Since the contract award, Continental has not sold its 420-B transmitter to any other broadcaster, which is a great pity. Brown Boveri, on the other hand has sold over 50 of its PSM transmitters to nearly 20 international broadcasting agencies, proving that these respect Brown Boveri transmitter technology. At the time of writing this book Marconi is thought not to have sold its B-6128 SW transmitter to other than the VOA and the BBC. However, the company is confidence that it will succeed in securing the outstanding options in the VOA contract.

28.2 VOA tender

In essence the tender was for the upgrading of the VOA network to make the Voice of America the largest and most powerful international broadcasting organisation in the world. The Request For Proposals (RFP) was originally for a hundred 500 kW transmitters, but this was subsequently reduced to 55. Even at this reduced figure it represented the largest and most ambitious broadcasting project in the history of broadcasting.

It called for new shortwave relay stations in Morocco, Thailand, Sri Lanka, Botswana, Puerto Rico (option) and Israel (option). In addition to this the tender called for antenna subsystems, satellite connect subsystems, network control subsystems, PBX subsystem, audio/communications subsystem and power plant subsystem. The plan for the transmitters is given in Table 28.1.

Table 28.1 *VOA expansion programme*

Botswana	4 × 500 kW SW	2 × 600 kW MW
Morocco	10 × 500 kW SW	
Puerto Rico	9 × 500 kW SW	
Israel	16 × 500 kW SW	
Thailand	7 × 500 kW SW	
Sri Lanka	7 × 500 kW SW	
Total 55 transmitters, 53 SW 500 kW, and 2 MW 600 kW.		

Chapter 29

Commercial giants: French broadcasting

The early history of French broadcasting, which predates that of many other European countries, was briefly described in some earlier chapters. The way it has evolved over the decades, with its distinctive blend of state-controlled broadcasting working with commercial broadcasting, deserves some particular attention.

29.1 Infrastructure

TeleDiffusion de France (TDF), with its registered office at Montrouge, is responsible for the operation of radio and television transmitters. Radiotelevision Française d'Outre-mer (RFO) with its headquarters in Paris, is responsible for the control of broadcasting in the French Overseas Territories. Radio France (RF) also based in Paris, is the programme-originating authority for broadcasting. At the same location, but with a different administration, is Radio France International (RFI), the controlling body and the originator of programmes for foreign broadcasting.

The French state operates 37 AM broadcasting stations in the long and medium wavebands with carrier powers ranging up to Allouis, which broadcasts 2000 kW on long waves. At least 16 of the radio stations have carrier powers between 100 and 600 kW. Excluding the low power (less than 100 kW) stations, the total broadcasting capacity comes to 4200 kW. The State also operates some 90 FM stations. Both networks radiate a number of programmes ranging from popular music to education.

RFI has access to some of the high-power MW transmitters, which it uses to broadcast for foreign workers, which go out in Arabic, Cambodian, Laotian, Portuguese, Serbo-Croat, Spanish, Turkish, Vietnamese and French for Africans. For overseas broadcasting, RFI has its main SW transmission complex in central France. Here it has three separate stations no more than a few kilometres apart. Allouis is fitted with four 100 kW transmitters, Issoudun C with eight 100 kW transmitters and Issoudun E with eight 500 kW transmitters. When these stations were built, they represented one of the largest concentrations of SW transmitters in the world.

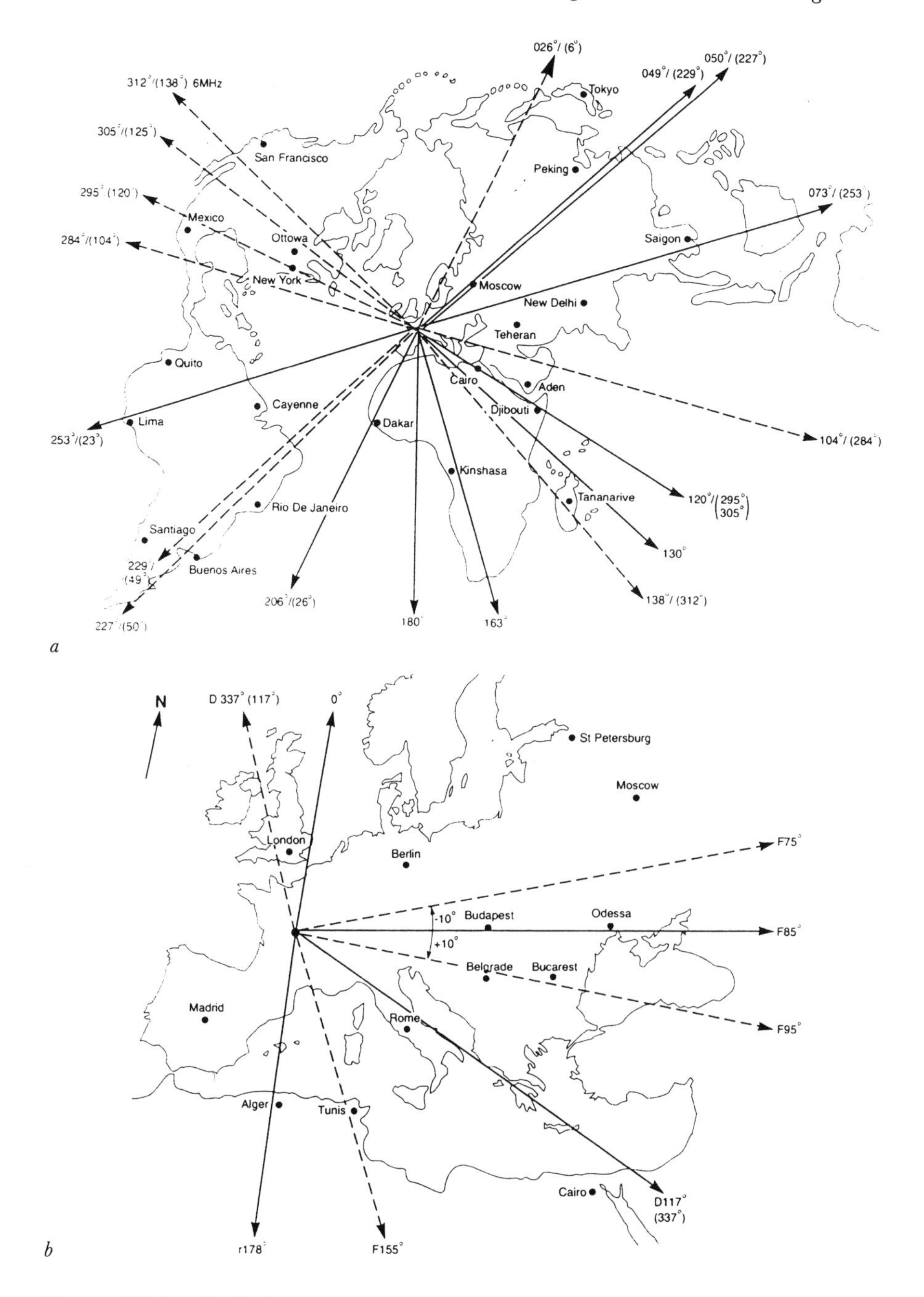

Figure 29.1 *Radio France International: network coverage*
Angles of fire from short wave transmitter stations
a Issoudun C
b Allouis

Figure 29.2 *SW station at Issoudun, France*

Additionally, RFI has two main relay stations elsewhere in the world. At Moyabi, Gabon, there are four 500 kW transmitters, and at Montsinery in French Guiana are four 500 kW transmitters. In Cyprus, RFI broadcasts over a 600 kW MW transmitter on a frequency of 1233 kHz. Radio Monte Carlo Middle East is a separately owned radio station, which leases its transmitter to RFI, Trans World Radio (TWR) and Radio Monte Carlo at different times of the day.

This is where distinctions between state-owned broadcasting authorities and commercial broadcasting organisations become a little blurred. But then, RFI is not the only international broadcasting authority to make use of the near-perfect strategic location that Cyprus offers. The BBC World Service has a 1000 kW capability for MW broadcasting on the same island.

As to French commercial broadcasting, in 1988 a total of 1669 licences had been issued. Not all of these are operational, but over 600 are affiliated to about ten private commercial networks. One of the most important privately owned commercial stations is Sud Radio, Toulouse, which operates three MW and 14 FM stations. Europe 1, another private commercial station, operates 30 FM stations. Its main station, however, is located at Felsberg in the Black Forest. This is a giant 2000 kW LW transmitter, which broadcasts on a frequency of 183 kHz. This radio station serves the whole of France and Switzerland, and much of western Germany. One of the world's largest broadcasting stations, it ranks as an international broadcaster, but is run on purely commercial lines.

Radio Tele Luxembourg (RTL) is another commercial broadcasting company, based at Luxembourg where it has been since the 1930s not counting a period of four years when it was under the control of German army technicians. RTL operates about 30 FM stations in French cities and towns, but its high power AM transmitters are in Luxembourg. On Juglinster Plain it has a 2000 kW LW transmitter broadcasting Radio Luxembourg on a frequency of 234 kHz, and one 500 kW and one 10 kW SW transmitter. At Marnach it has another high-power transmitter, of 1200 kW carrier power, on a frequency of 1440 kHz (medium wave).

Finally, to complete the three-ring circus that is the largest and most powerful collection of super transmitters in Europe, there is Radio Monte Carlo (RMC). This is easily the biggest commercial broadcasting company in Europe. The name is misleading; it conceals the identity of a complex network, which broadcasts on every part of the frequency spectrum, from long waves to VHF. RMC has three main AM transmission centres: Fontbonne, La Madone, and Roumoules, all situated on the lower slopes of the French Alps, overlooking the Mediterranean. RMC is not only the largest commercial broadcaster in Europe; it has an unusually fascinating history, of which a summary is given later in this chapter.

29.2 Technology

Underlying the many different broadcasting authorities, whether state controlled or driven by private capital, is French broadcasting technology. France is the only country in Europe, and one of only two in the world, capable of complete self-sufficiency in the design and manufacture of broadcast equipment. From audio/video to radio frequencies, from studios to transmitters to transmission systems, it is French equipment that dominates the broadcasting scene. In the main this equipment comes from one of the many divisions or companies within the state-owned Thomson SA.

Inventiveness is a characteristic of the French. French scientists and engineers were laying the foundation stones of super-power broadcasting in the late 1930s, in the shape of triodes that could handle hundreds of kilowatts of power. During the Nazi occupation of France, after June 1940, the research work was concealed from the occupying forces. Eventually the results of this work, together with a prototype tube were spirited away to America. From this first sample, US companies were able to produce the Machlette triode, which developed 250 kW of radio frequency power.

After the war, France resumed this research programme and soon regained world supremacy in vacuum-tube power-grid technology. This is illustrated in the tetrode TH 539 which powers 2000 kW transmitters for LW/MW broadcasting, and the TH 558 which powers 500 kW SW transmitters, manufactured by Thomson-CSF and used by most other manufacturers of these equipments.

29.3 French foreign policy and international broadcasting

Information (propaganda) broadcasting can take many forms; the common underlying factor is that all reflect the foreign policy of the country originating the broadcasting, although this is sometimes denied by the broadcasting authority (the BBC World Service is one example). The purpose of propaganda can range from the export of culture, information and disinformation broadcasting, evangelism, advertising or politics.

Figure 29.3 *Advertisement in WRTH for Radio France International*

Numbers of different languages broadcast is a parameter often used by international broadcasting authorities; the BBC, for example, broadcasts in 36 languages, 24 hours a day over 87 transmitters, and politics features strongly in its output. Measured by such parameters, RFI is well down the world table of international broadcasters. RFI broadcasts 90% of its total output in French, or dialects of French.

RFI is basically culturally, rather than politically, driven. It did not engage in the Cold War and the crossfire of words between East and West. Significantly, RFI has never suffered from Soviet jamming, unlike the BBC, VOA, Deutsche Welle and Kol Israel, and France does not permit any other country to station its propaganda transmitters on French territory (unlike the UK, Germany and Israel).

It is largely due to the wisdom and foresight of de Gaulle himself that RFI, along with commercial broadcasters such as Radio Monte Carlo, pursues a policy of culture-driven broadcasting. It was he who realised the potential of foreign broadcasting: much of the French colonial empire was strung out along North Africa, and in Central and West Africa. RFI caters for these countries, along with other French-speaking countries in the Middle East, the Caribbean and parts of the Pacific. Perhaps one tangible proof of the success that France is obtaining with its policy on international broadcasting is the export of French goods. In the world of international broadcasting, it is significant to note that more than 30 different countries in Africa alone have invested in French broadcasting equipment, and one Middle East state possesses the largest SW station in the world, based entirely on French technology.

29.4 Radio Monte Carlo

Although it is registered as a private company, Radio Monte Carlo is indirectly owned by the government of France in partnership with the principality of Monaco. It also happens to be one of Europe's largest broadcasting organisations, and one of the most influential in the field of commercial broadcasting.

Ironically, RMC owes its origins to the Third Reich. In 1942 the Germans built an underground bunker on Mont Agel, Monaco, high above the Mediterranean coastline. The project was part of a long-term plan to broadcast radio propaganda during the thousand-year Reich. It was an excellent choice for a transmitter site, but the Germans never had time to finish it before the collapse of the Third Reich.

The French, on the other hand, took over the bunker along with dozens more on the Atlantic and Mediterranean coastlines, and during the liberation of France the resistance began to broadcast radio messages from Monaco, calling itself 'Maison de la Radio'. This was the birth of Radio Monte Carlo.

RMC is only one of a number of broadcasting agencies in France, and although it is advertising-driven (unlike others such as RF and RFI) RMC is important in the broadcasting world for its projection of French culture in Italy, the Near East, and particularly to France's erstwhile colonies in North and Central Africa: Tunisia, Morocco, Algeria, Chad and others.

RMC is not a single radio station: it broadcasts on all wavebands, from LW to VHF-FM. It has studios in Monaco and at Rue Magellan – in the most fashionable part of Paris, connected by a satellite link. The main transmission relay centres

Figure 29.4 *Radio Monte Carlo. View of transmission line 2 MW rating*

Figure 29.5 *Antenna switch gear at SW station, Issoudun, France*

are at Fontbonne on Mont Agel, the original Nazi site—a SW complex with two 100 kW and one 500 kW transmitters, with ten high-gain curtain arrays. La Madone, on the western end of the col, is a MW transmitter site equipped with two 600 kW transmitters in a coupled arrangement to produce 1200 kW. This station radiates programmes on 702 kHz during the day for RMC. At night it switches to 1467 kHz for broadcasting skywave long-distance service for Trans-World Radio.

The third high-power AM transmitting station is at Roumoules, on a flat plateau on the lower slopes of Les Hautes Alpes. Roumoules is one of the two stations in the world with the capability of being configured for 3000 kW carrier power. At present it radiates 2000 kW on the LW band at a frequency of 216 kHz carrying commercial radio programmes for RMC, which can be received all over Western Europe. Also at Roumoules is a new state-of-the art 1000 kW MW installation, coupled to a very advanced design antenna system. Its uniqueness is in its ability to radiate on either of two frequencies: 702 kHz by day and 1467 kHz by night, and with a polar diagram that can be switched to any one of four azimuth bearings to suit the target zone.

Because of the ideal characteristics of the Roumoules site—a large flat plateau with excellent Fresnel zone qualities and extremely high electrical conductivity—the antenna system is able to provide a gain of the order of 8.5 dB relative to a dipole on three of the bearings, and a lesser gain on the fourth. With this large amount of gain, added to 1000 kW of power, this transmitter is capable of blanketing well into Russia as far as the Ural Mountains or to other regions of Eastern Europe, depending on the selected lobe. The azimuth bearings have a beam width of ±32° at the 3 dB points, and switching time is 4 s—incredibly fast, considering the massive size of the base tuning inductances at each of the five 100 m masts.

Table 29.1 *Sector coverage of TWR*

Azimuth bearing	Gain	Target zones
25°	8.6 dB	Berlin, eastern Germany, Poland, Russia west of Urals
85°	8.4 dB	Yugoslavia, Hungary, Romania, Bulgaria
240°	6.0 dB	Portugal, North Africa
320°	8.5 dB	UK, France, Scandinavia

When used in the above service, the transmitter is on leased service to Trans World Radio, a religious broadcasting organisation, and the programmes carry the identity of that name. In France, RMC is therefore better known for its FM programmes radiated over a network of transmitters in nearly 40 towns and cities in France and along the Côte d'Azur. The same programme is duplicated on LW and MW wavebands 1400 m (216 kHz) programme Français, 428 m (702 kHz) programme Italien, and 205 m (1467 kHz) programme Français. Catering for diverse tastes in music, RMC broadcasts from rock to classical.

Annecy, Avignon, Biarritz, Bordeaux, Dijon, Clermont Ferrand, Grenoble, Limoges, Lyon, Marseille, Aix, Montpellier, Nancy, Nice Saint Nazaire and

Toulon Valence are just a few of the FM stations, and in 1988 RMC extended its FM network to take in Paris. Originally located on the Champs Elysées, its transmitter is now installed on the Eiffel Tower. RMC uses a mixture of different manufacturers' FM equipment, including Thomson-LGT, Itelco and a few others. At low power, VHF-FM equipment is a very competitive market, with a wide range of manufacturers to select from. For megawatt broadcasting on AM wavebands it is a different story, as there are only about three companies with the necessary skill and experience at the level required to execute the design and construction of megawatt broadcasting complexes.

Naturally, when RMC first contemplated the building of a LW transmitter in 1974 it turned to Thomson-CSF, but its relationship with the company extends much further back. All of the SW transmitters at Fontbonne are of Thomson manufacture. The design and construction of transmitting stations such as Roumoules is a very advanced science, made possible by a massive continuing investment in transmission technology.

The task of designing the massive antenna systems has been made possible with the advent of high speed computers and extensive development of optimisation computer programmes, which have made modelling feasible. Today, using such techniques, complex equations in electromagnetic propagation can be solved in a fraction of the time that would have been required two decades ago. With a certainty never before possible, it is now possible to predict the performance of closely coupled antenna systems, such as the LW and MW systems at Roumoules, without resorting to extensive testing with scaled model layouts.

The design and construction of megawatt transmitters such as those at Roumoules is the product of 50 years of experience in transmitter design and investment in tube technololgy. The installation at Roumoules is an indication

Figure 29.6 *Antenna matching unit at Radio Monte Carlo*

of the attention now being given by broadcasting agencies to those AM wavebands once reserved for national and regional broadcasting, as a medium for international broadcasting. This trend is set to continue. Several years ago, many companies were building high-power AM transmitters, but expanding technologies in multi-disciplinary fields has left some companies by the wayside. Today, practically all major projects in super-power AM are being executed by one of two European companies: Asea Brown Boveri and Thomson-CSF. Perhaps the best indication yet of the new growth market for mega-powered installations is the largest ever AM transmitter complex in the history of radio broadcasting; three 2000 kW transmitters for MW broadcasting, completed by Thomson-CSF in Iraq, in 1986.

Table 29.2 *The history of Radio Monte Carlo*

1942	As part of the plan for the 'Thousand Year Reich', the Nazis decide to construct a propaganda radio station, with its studios in Monaco and the transmitters in an underground bunker at Fontbonne, Mont Agel.
1945	After the liberation of France the French Government and Monaco take over the company, and create Radio Monte Carlo.
1946	First programmes broadcast on a small transmitter from 'Maison de la Radio'. A 10 kW MW transmitter was installed shortly afterwards, rapidly followed by two 30 kW SW.
1949	120 kW MW transmitter installed, with a new antenna.
1954	200 kW MW transmitter installed, followed soon afterwards by another, giving a transmitted power of 400 kW.
1960	Additional transmitter site built at Col de la Madone, with two 100 kW SW transmitters and seven directional antennas.
1964	RMC start up frequency modulated broadcasting on 98.5 MHz.
1965	RMC builds its first LW transmitter, a 1200 kW super transmitter—the first in Europe. This is also located at Col de la Madone, and a new three-mast LW directional antenna, 320 m high.
1965	New offices and studios established in Paris, Rue Magellan.
1974	State-of-the-art LW station constructed in Roumoules in the Haute-Provence Alpes. With this powerful 2000 kW transmitter and new three mast antenna the primary range of service is tripled.
1974	Daytime primary service now reaches Italy on 702 kHz. At night the frequency is changed to 1467 kHz, leased to Trans World Radio.
1978	Birth of RMC Côte d'Azur service in stereo from FM stations.
1979	A new standby LW antenna erected to provide better service reliability in electrical storms prevalent in the Alpes.
1981	Two new FM programmes, RMC Rock and RMC Classique, for Nice, Cannes et Côte d'Azur, Monaco and Menton.
1982	Fontbonne receives a new 500 kW SW transmitter, and a six-direction antenna system.
1984	Transmitter building at Roumoules extended to allow the installation of a 1000 kW standby transmitter for LW use. With this additional unit Roumoules has 3000 kW installed capacity with 2000 kW in regular service. The LW antenna system is upgraded to 3000 kW rating, to cater for future expansion.

Table 29.2 *Continued*

1984	50 kW MW transmitter installed at La Madone as an additional service while new 1000 kW transmitter under construction at Roumoules.
1987	New state-of-the-art 1000 kW MW transmitter installed at Roumoules, together with a four-directional antenna to allow transmission on 325°, 241°, 85° and 25° azimuth. This 1000 kW transmitter can blanket from Scandinavia to west of the Ural Mountains and as far south as Algeria and Tunisia—870 million people in 38 countries.
1988	RMC broadcasts FM from the Eiffel Tower. By the end of 1988 thirty French towns can receive RMC FM.
1989	Through the Roumoules 1000 kW MW transmitter, Trans World Radio broadcasts in Albanian, Arabic, Armenian, Berber, Catalan, Croatian, Czech, Danish, English, Estonian, Farose, Farsi, French, German, Greek, Hebrew, Hungarian, Italian, Kabyle, Kazakh, Kirgiz, Kurdish, Latvian, Lithuanian, Macedonian, Norwegian, Polish, Romanian, Russian, Serbian, Slovak, Slovenian, Spanish, Turkish, Ukrainian and White Russian.
	These countries are covered by night time broadcasting; 2000-2400 hours and 0500-0700 taking advantage of ionospheric reflected waves.

Chapter 30

Super power in the Arab world

Although Western and Arab civilisation evolved alongside one another, the social, economic, political and technological upheavals in Europe from the 18th Century onwards left the Arab world largely untouched. The first time the Arab world came into contact with many products of 20th Century technology was as a result of becoming a battlefield for European powers in the Second World War. Although suspicious to begin with, the people of these countries soon took to radio technology like a duck to water. The reasons for this behaviour are to be found in the countries themselves.

The Arab countries of the Middle East and North Africa are linked by historical, geographical, climatic, economic and linguistic features, but most importantly by the Islamic faith. The Arab culture, by long tradition, is orally based, where emotion and rhetoric are regarded more highly than structured argument. Moreover, the spoken word predates the written by several thousand years, and is still more highly regarded.

Thus radio broadcasting was a natural medium for these countries. The radio receiver became a social instrument—no restaurant or cafe in the towns and cities was complete without one, and in the tea houses of the villages, the *fellaheen* would gather together after a day's toil to relax, hear and be entertained by voices and music.

From the early 1950s, these countries wanted to have broadcasting stations but there were two major problems: the cost, and the fact that the equipment would have to be imported from the USA or Europe. In the end the equipments were supplied on very favourable terms, by the Americans, British or Soviets, all of whom sought something in exchange, usually military bases. In other cases, the terms gave the occupying powers the right to build their own radio stations for propaganda broadcasting purposes, and in some instances the right to access the country's transmitters with their own programme feed, again for propaganda purposes. Thus the Arab countries learned there is no such thing as a free lunch.

The next milestone in Arabic broadcasting was the transistor revolution. Possibly because of the lack of indigenous suppliers, these countries became the first target for Japanese mass-production methods. Over a million of the first-generation transistorised AM receivers were dumped at low prices. Now, for the first time in history, people in small villages and even nomadic tribes could hear radio broadcasts while on the move.

From this point it was but a short step to the political figures of the Arab world realising that they had within their grasp a new form of mass communication, which, by its nature as an oral tool, was uniquely suited to Arab culture; an instrument with which they could communicate directly to the peoples in the towns and villages, without any intermediary, and more importantly, without the fear of being censored or intercepted. Thus the transistor revolution was the precursor to another kind of revolution: the overthrow of many old regimes.

In recognition of the importance of radio broadcasting—which in Saudi Arabia had been pronounced by religious leaders some 20 years earlier as the work of the devil—all these countries began to increase their investment in broadcasting power, such that by the mid 1960s many had transmitters with powers of 250 kW. In lessons learned from colonial rule, the Arab countries showed an awareness that radio was much more than a social instrument: it was a strategic weapon, which should remain firmly under government control. To this end, practically all the Arab states passed decrees to establish government control, with technical responsibility vested in a PTT and exploitation in ministries of information. Wealth from oil revenues provided the means of funding expansion in broadcasting power which today dwarfs the broadcasting capability of many countries in the West.

An abundance of oil did more than create wealth for expansion in radio broadcasting: the energy surplus meant that, while some Western countries cut back on broadcasting power because of energy costs, the Arab nations had no such problems. The fact that a megawatt transmitter demands 20 tonnes of oil per day to generate electrical power was of no consequence.

When technology made it possible to design and construct transmitters with outputs of 1000 kW carrier power, the Arab countries were the first to invest in them (apart from the USA, which funded the purchase of a small number by Voice of America). To the Arab countries, broadcasting with megawatt power made it possible to target neighbouring Arab countries with their national broadcasting systems. Although regional variations are quite strong, Arabic is, broadly speaking, a common language from Morocco in the west to Oman in the east, from Syria in the north to Sudan in the south. This meant that the Arab nations needed to draw no sharp distinction between national and international broadcasting. A typical Arabic programme coming out of Cairo studios, broadcast from the 2000 kW MW transmitter at Batra, can be enjoyed equally in Tunisia to the west or Saudi Arabia to the east. To this extent, Arab broadcasting is less hypocritical than that practised in some countries in the West. The high quality of programmes broadcast by the BBC World Service contrasts sharply in style, content and presentation with the standard broadcast over the BBC's domestic networks Radio 1 and 2. Even so, some Arab countries do broadcast specially produced programmes for a neighbouring country.

The first Arab country to invest in super power technology was the Kingdom of Saudi Arabia, the most conservative and autocratic of all the Arab states. Under American influence, negotiations between the two governments resulted in a contract for the Dallas-based company Continental Electronics to supply two 1000 kW transmitters and the associated antennas and diesel generators early in 1970, at Qurayat and Jeddah. These were followed by five 2000 kW systems for installation at Duba, Qurayat, Jeddah and Damman. Thus, by the late 1970s,

Saudi Arabia had acquired 12 000 kW of broadcasting power. At about the same time, Continental Electronics supplied 2000 kW transmitters to both Egypt and Jordan, installed at Batra and Aijun.

Also actively engaged in constructing super transmitters at that time were Brown Boveri, Thomson-CSF, the Czech company Tesla, and Radio Industries Zagreb of Yugoslavia, which had a marketing agreement with Siemens of Berlin. Libya acquired two 1000 kW, and one 2000 kW transmitters from RIZ between 1974 and 1978. Syria bought from both Tesla and RIZ, the Tesla system being a 1500 kW transmitter, and a 1000 kW RIZ transmitter installed at Damascus.

From the late 1960s to the late 1970s, Thomson-CSF supplied six 1200 kW transmitters to Iran, Iraq and Saudi Arabia. Most of the seven 1000 kW transmitters built by Brown Boveri during the same period went to Arab countries. Practically all the Arab countries also have significant numbers of lower powered transmitters, most of which were installed before the 1000 kW transmitter era of the mid 1970s onwards.

Because of the high capital investment that a super-power transmitter represents, they tend to remain in service for anything up to 40 years, and sometimes even longer, after which they are relegated to a standby role, with a corresponding cut in power in deference to old age. Thus, many of the transmitters supplied in the 1950s are still in service. Even so, the sales of AM transmitters for MW and LW broadcasting service continue to grow with 1000/2000 kW transmitters becoming more popular. From super power, 1000 kW has become the norm.

In 1982 Thomson-CSF supplied a 2000 kW LW system to Radio Mediterranée Internationale, the government-owned broadcaster of Morocco. Three years later, Thomson supplied three 2000 kW systems to a Middle East state for MW operation. Within the past five years, ABB has supplied a system to Jordan including a 1000 kW MW transmitter and a 1200 kW LW transmitter, comprising part of the Qasr Kherane project. More recently, the same company has commissioned two 1200 kW MW systems to Islamic Republic of Iran Broadcasting (IRIB).

One of the largest AM projects ever undertaken rivalling the three 2000 kW systems mentioned earlier, is the Algerian LW project. ABB has supplied to Radiodiffusion-Télévision Algerienne two 2000 kW LW transmitter systems, installed at Ouargla and Bechar. These two sites are 1900 km apart, on the northern edge of the Sahara, south of the Atlas Mountains. Each system feeds 2000 kW of carrier power into a three-mast, high-gain inline array. Since the stations were handed over, both have performed without interruption, broadcasting in a parallel operation on two LW frequencies—a massive 4000 kW of carrier power serving the huge land mass of Algeria, with minimum power radiated in the northerly direction.

30.1 SW broadcasting

Arab interest in the use of the high-frequency spectrum did not kindle until the late 1950s. From a slow start, the pace accelerated as oil revenues began to build up and the oil-producing states wanted to have a voice in international affairs. The greatest acquisition rate in SW technology took place after the first 500 kW SW transmitters were developed by three companies at about the same time, but more particularly since the advent of the high-efficiency PDM system by AEG in 1982, applied to the 500 kW level of power.

Tables 30.1 and 30.2 give sales figures to Arab countries and Iran, extracted from the lists of total sales by the companies named. It should be stressed that these figures are not an indication of the total sales. AEG, for example, is strong in sales to European broadcasting authorities, whereas ABB is particularly strong in North Africa and the Middle East, as is Thomson-CSF.

Table 30.1 *Recent sales of LW/MW super transmitters to Arab countries*

Country	Location		Supplier	Year
Libya	MOI		Radio Industries Zagreb	1978
Saudi Arabia	MOI		Continental	1979
Saudi Arabia	MOI	Duba	Continental	1979
Saudi Arabia	MOI	Qurayat	Continental	1979
Saudi Arabia	MOI	Jeddah	Continental	1979
Saudi Arabia	MOI	Damman	Continental	1979
Egypt	OBTF	Batra	Continental	1979
Jordan	HBS	Aljun	Continental	1979
Morocco	RMI	Nador	Thomson-CSF	1982
Abu Dhabi	MOI	Dabia	Asea Brown Boveri	1985
Algeria	RTA	Bechar	Asea Brown Boveri	1986
Algeria	RTA	Ouarglia	Asea Brown Boveri	1986
Iraq		Tanaf	Thomson-CSF	1986
Iraq		Maisan	Thomson-CSF	1986
Iraq		Sulaimaniya	Thomson-CSF	1986

Table 30.2 *Recent sales of 500 kW SW transmitters to Arab countries*

Country	Site	Transmitters	Supplier	Year
Libya	Sabratha	4 × 500 kW	TH-CSF	1977
Syria	Damas	4 × 500 kW	TH-CSF	1983
Kuwait	Kabd	4 × 500 kW	ABB	1980
Iraq	Babylone	2 × 500 kW	TH-CSF	1983
Iraq	Balad	16 × 500 kW	TH-CSF	1985
UAE Abu Dhabi	Dabyia	4 × 500 kW	ABB	1985
Jordan	JRT Qasr Kherane	3 × 500 kW	ABB	1988
Kuwait	MOI Kabd	2 × 500 kW	ABB	1988
Kuwait	MOI Kabd	4 × 500 kW		
Iran	MOI IRIB	2 × 500 kW	ABB	1989/90
		4 × 500 kW	ABB	1990/92
Kuwait	Kabd	4 × 500 kW	ABB	1990
Saudi Arabia	MOI Riyadh	4 × 500 kW	TH-CSF	1990
Iran	IRIB Sirjan	10 × 500 kW	AEG	1990/93

The sales figures for AEG refer to its PDM range, those for ABB to its PSM range. Thomson-CSF introduced its PDM range later than the other companies, so that most of its sales refer to 500 kW transmitters fitted with Class B modulation.

From this list of sales of super transmitters to Arab countries, the trend is quite apparent: after a short pause after the acquisition of MW transmitters the Arab countries moved to the acquisition of 500 kW SW transmitters to extend the voice of Islam to the world.

Table 30.3 *Broadcasting power in kW of leading Arab countries*

			Broadcasting power in kW		
Country	Population (millions)	Receivers (millions)	MW/LW	SW	Total
Iraq*	17	3.2	9400	9000	18 400
Saudi	12.4	3.9	12 800	4500	17 300
Iran	49.5	11.0	8800	7000	15 800
Libya	4.5	1.0	5800	2000	7800
UAE	1.4	0.37	3250	4200	7450
Kuwait*	1.9	1.1	3000	7500	10 500
Jordan	2.9	0.7	4200	2000	6200
Morocco	25.1	4.3	5300	850	6150
Egypt	53.3	13.6	4000	700	4700
Syria	11.9	2.5	2500	2000	4500
Algeria	24.6	5.4	7300	300	7600

*before August 1990

The figures in Table 30.3 do not necessarily agree with published data in the WRTV Year Book, as they are my own estimates based upon different sources and include known orders for delivery up to 1991. They include high-power transmitters only and take no account of low-power transmitters in service. Iran is not strictly an Arab country, but has been included because of its geographical and cultural closeness to Arab countries, and its importance in Middle East affairs.

A comparison of broadcasting power throws up some interesting facts. Saudi Arabia is still the most powerful broadcaster on medium wave and is presently building up a greater capability in short wave. Iraq (until its capacity was devastated in the Gulf War) was the most powerful voice on all wavebands, followed by Iran, which was set to overtake Iraq by 1994, when it will have ten more 500 kW SW transmitters in service.

Kuwait—also, until the Gulf War—was in a class of its own: the smallest of the OPEC states in population—considerably less than many European cities—yet with one of the greatest figures for GNP per capita in the world. As befits such a wealthy state, it was developing its broadcasting capability to match, so that on a per capita basis it was the world's most powerful broadcaster. Falling into the same category is the United Arab Emirates. The figures for UAE include the 'Voice of the United Arab Emirates', broadcasting on medium and short wave, and UAE radio and television stations in Dubai on medium and short wave.

Recently Abu Dhabi acquired powerful transmitters from ABB for MW and SW broadcasting.

Egypt was the first Arab country to embrace radio broadcasting, and is the most populous of all Arab nations. Its receiver count is still the highest of all Middle East and North African countries. Nowhere is the radio receiver more in evidence than in Cairo, but Egypt is being left behind in the race for broadcasting super power by the more wealthy OPEC states. Egypt still operates one 2000 kW MW transmitter at Batra, with a number of smaller powered MW stations, up to 600 kW.

The Hashemite Kingdom of Jordan, one of the smallest of the Arab states, with a population of only 2.9 million, has set a standard in broadcasting technology which will probably become the model for the future. In April 1988 a new broadcasting complex at Qasr Kherane was opened by King Hussein. 70 km from Amman in the Jordanian desert, on the site of an ancient fortress, ABB has built a multi-frequency, computer-controlled transmitting station, comprising three 500 kW SW transmitters, one 1000 kW MW transmitter and two 600 kW LW transmitters. Qasr Kherane is not only the first multi-frequency broadcasting station in the world, with LW, MW and SW transmitters within the same building; it is also the most powerful SW system with two of the 500 kW transmitters feeding into one curtain array, producing 1000 kW of carrier power. The gain and directivity of the HR 8/4 curtain array means that Jordan can now project its SW transmissions over 15 000 km, requiring six hops, realising a field strength of 50 dBμ.

Having built up a super-power capability with which to broadcast to the Islamic world, the Arab nations have sought to have a stronger voice on the short waves to exert a greater influence in world affairs. The next phase is the quest for technical superiority in world broadcasting: multi-frequency, global broadcasting capability

Figure 30.1 *Transmitting centre erected by Asea Brown Boveri (ABB) at Qasr Kherane in Jordan. It comprises three 500 kW short wave transmitters, one 1000 kW medium wave transmitter and two 600 kW long wave transmitters*

Figure 30.2 *Transmitter station power control room at Qasr Kherane, Jordan*

is now the goal. In many Western countries, financial constraints make it necessary to adapt or extend existing transmission stations; Middle East countries are under no such constraints. Adequate funding enables broadcasting authorities to demand latest technology, coupled to super power and a high order of transmission reliability. Even with high RF powers, the transmitter stations must be capable of full remote control and running in a semi-attended or even unattended mode.

Practically all of these new stations are located in barren, hostile deserts of the Middle East, one of the most inhospitable terrains of the world, characterised by temperatures of 50°C in summer and down to −40°C at night in winter. Altitudes of 5000 feet bring blizzards and 1–2 metres of snow, and spring brings severe flooding. To carry out a major installation project under such environmental conditions, the broadcasting authorities prefer a contractor with a 'turnkey' capability to complete the entire project, including access roads, power-generating plants, and all main services including accommodation for staff.

The design and construction of the transmission centre itself, which includes transmitters, switching matrices, baluns, feeders, slewing switches and antennas, calls for a multi-disciplinary team of engineers with diverse skills, including the use of powerful computers for analysing the mechanical and electromagnetic properties of the many antenna systems, and the possible interaction problems from close proximity.

Advancing technologies, coupled with the need for a high investment by the contractor and the ever-present risk of failure due to some unforeseen hazard, has reduced the numbers of eligible turnkey contractors so that only the colossus companies have the required ability. As a result, all major broadcasting projects have been executed by either Thomson-CSF, ABB or AEG-Telefunken. There is little evidence of any British company being involved on anything like the scale of these broadcasting projects. This appears to be yet another valuable export industry in which Britain is unable to compete.

Recently, I carried out a survey of the sales of SW transmitters by the big four European companies: ABB, AEG, Marconi and Thomson-CSF. This analysis is based on world sales over the past four decades, since 1950. The remarkable growth in sales that has taken place in the 1980s comes mainly from orders received from the Arab countries. It should also be born in mind that these statistics refer to numbers of transmitters sold. If the trend in transmitter power output is taken into account then the growth rate is even more impressive. The high sales volume for ABB transmitters since 1985 coincides with the introduction of advanced PSM technology by ABB. (See Table 32.5).

Table 30.4 *Recent major broadcasting projects*

Country	Year	Project	Company
Morocco	1982	2000 kW LW system with a three-mast inline directional antenna, 5.6 dB gain	Thomson-CSF
Iraq	1984	2000 kW MW system with a three-mast inline directional antenna, 5.6 dB gain	Thomson-CSF
Iraq	1985	Repeat installation at another site	Thomson-CSF
Iraq	1985	Repeat installation at a third site	Thomson-CSF
Iraq	1985	16 × 500 kW SW transmitters plus 96 curtain arrays, one rotatable, and a 1600 (expandable to 2000) crosspoint matrix	Thomson-CSF
Algeria	1985	2000 kW LW system with a three-mast inline directional antenna, 6.4 dBi gain	ABB
Algeria	1986	Repeat installation at second site 1900 km away	ABB
Jordan	1966	Multifrequency high-power system comprising 1000 kW MW system with three-mast radiator, 1200 kW LW system with two-mast radiator, 3 × 500 kW SW transmitters and 13 curtain arrays. Two transmitters are in parallel to give 1000 kW carrier power, the first of its type in the world	ABB
Iran	1990-93	10 × 500 kW SW transmitters and associated antennas and matrix. One of the largest single orders for 500 kW SW transmitters	AEG

30.2 Libya

Of all the Arab countries, Libya is most often portrayed as the arch-enemy of the West. Since the successful revolution on 1 September 1969, an image has been created in America and Western Europe of a dangerous state under a fanatical leader. Since propaganda broadcasting is the most popular method of projecting culture and political beliefs, it is interesting to examine Libya's capability in this area.

Libya, once known as Cyrenaca, a name of biblical origins, was once the breadbasket of the Roman empire. In more recent history it was a fashionable resort of the kings and queens of Europe, who relaxed in its beautiful city of Tripoli. Yet, at the time of its 'independence', when it was taken away from Italy after the Second World War, it was commonly viewed as one of the poorest nation states in the world. Today, only 20 years after securing its independence from another colonial power—Great Britain—it has emerged as one of the richer countries in the world, measured by GNP per capita.

During the years of British administration under the rule of the puppet King Idris, revenues from Libyan oil flowed into the coffers of the US and UK oil companies, swelling the exchange of these countries. Within Libya itself, there was almost no education; illiteracy ran at around 80% during the 1950s and 1960s.

On 1 September 1969, Muammar al Qaddafi successfully overthrew Idris and proclaimed Libya an independent Arab state. Qaddafi and his colleagues represented something new in Arab thinking: a new educated elite, well read in many Western languages—rare in the Arab world. Under Qaddafi's vigorous leadership—he was only 28 at the time of the revolution—the country has been transformed. For the first time, there is education up to university level, and the quality of life of the average Libyan has been lifted dramatically.

Radio broadcasting is the most dominant form of mass communication in Libya. Qaddafi's first move in the overthrow of colonial rule was to seize control of the radio stations, which until then had been controlled by the British and US forces. These radio transmitters were of low power, but they served their purpose, proving once again the power of radio broadcasting in a revolution. The primary development objectives after the revolution were agriculture, electrification, transport, hospitals, and above all housing. As a result, the media, radio broadcasting, television and newspapers did not figure prominently in national development.

Investment in radio transmitters for SW and MW broadcasting did not begin until four years after the September Revolution. The first acquisition was a pair of 100 kW SW transmitters, followed one year later by the purchase of a 600 kW and a 1000 kW MW transmitter. In 1977 Brown Boveri supplied four 500 kW SW transmitters.

In 1977–1978 another expansion phase took place when Libya acquired one 600 kW, one 1200 kW and one 2000 kW transmitter for MW broadcasting. All these were supplied by RIZ of Yugoslavia. At the same time it bought four of the latest 500 kW SW super transmitters from Thomson-CSF and after this it bought a 2000 kW MW transmitter from Brown Boveri. Thus, at different times, Libya has purchased transmitting equipment from several companies in Europe.

Practically every Libyan household today has at least one radio, and radio broadcasting is easily the dominant form of media in the country. Libyans listen to the radio while they drive to work, at work and at leisure.

In Libya, as in the Arab world generally, there is no sharp distinction between the types of radio programmes for national and international listening audiences. From the powerful MW transmitters in Tripoli and at other sites, programmes are transmitted over considerable distances after the hours of darkness. Listeners in adjacent Tunisia, Algeria, Egypt and Sudan are treated exactly like listeners in Libya itself. This approach to a unified programme standard is typical of almost all Islamic countries concentrated along the North African coast and eastwards as far as Iraq.

Chapter 31

Religious broadcasting and propaganda

Purely political propaganda continues in importance, but it has a growing rival: religious broadcasting. The first religious faith to embrace the newly discovered short waves was the Roman Catholic Church. The first experimental transmission took place in Italy; the transmitter had been constructed and installed by the Marconi Company, under the supervision of Guglielmo Marconi himself. This was a momentous occasion for Pope Augustus, and an appropriate one, in that it was an Italian who had given the world this new window on the universe.

Since that first-ever experimental SW transmission, the Vatican State has expanded and regularly updated its radio transmitters such that it can claim to have one of the most modern broadcasting centres in the world. From its studios in the Vatican City, programmes are carried by landline and microwave to the transmitter sites. Within the Vatican City itself are low-power AM and FM transmitters, and the high power SW transmitter complex is located at Santa Maria di Galearia, outside the boundary of Vatican City. The Vatican City is still the world's largest religious SW broadcaster, with three 500 kW SW transmitters, two of Thomson-CSF manufacture and one from AEG. The Vatican City is one of the few broadcasting agencies equipped with a 500 kW rotatable curtain array, also of AEG manufacture.

For many years the Vatican City was the only religious broadcaster on the short waves. In the early 1930s, however, the Worldwide Broadcasting Foundation, one of six private broadcasting companies licensed to operate on the short waves, began to broadcast religious programmes over WRUL, a 50 kW transmitter at Scituate, Massachusetts. Also operating in the USA from the early 1930s was W1XAL, a quasi-religious broadcaster with the slogan 'Dedicated to Enlightenment'.

During the war, all privately owned and operated SW stations were leased by the US Government to carry VOA services, and this wartime arrangement went on right up to the early 1960s, when VOA brought its own transmitters into service. When this arrangement finally terminated, three of these stations (WRUL, KGEI, Redwood City, California, and W1NB, at Red Lyon, Pennsylvania) went over to religious broadcasting.

After 1963 the US Federal Communications Commission, complying with the Radio Regulations of the ITU, stopped issuing new licences to privately owned stations. This freeze remained in operation until new rules came into place, in July 1973, heralding the renaissance of private SW broadcasting in the USA. From then on the numbers of religious stations operating in the USA increased dramatically.

Meanwhile, other religious broadcasters began to appear in other regions of the world. Trans World Radio was founded by Dr Paul E. Freed in 1952, as International Evangelism Inc., for mass communication of the gospel message. Today, TWR has grown from its first 2500 W transmitter, which came into service on 22 February 1954 in Tangier, to one of the largest broadcasting organisations in the world. TWR now broadcasts in more than 75 languages; its own press statement ranks it alongside Radio Moscow, as 'one of the largest radio broadcasting operations of any kind, religious or secular'. In fact, a comparison with the operations of VOA might be more accurate—TWR, like VOA and unlike Radio Moscow, has its transmitters located in every continent of the world: Monte Carlo, Cyprus, Swaziland, Sri Lanka, Guam and Montevideo, Uruguay. This last station by itself serves over 11 million people. Also in the Western hemisphere it has Radio Bonaire in the Antilles, a 500 kW MW transmitter. The most powerful of all its transmitters is a 1000 kW MW transmitter at Roumoules, near Marseilles, leased for night-time broadcasts to Eastern Europe; the Roumoules transmitter like the 600 kW MW transmitter on Cyprus, is owned by Radio Monte Carlo.

Although TWR has no transmitter sites in the USA its headquarters is at Chatham, New Jersey. Another religious broadcaster association with no transmitter sites but other connections in the USA is the Far East Broadcasting Association. FEBA has its registered office in the south of England, at Worthing, Sussex, and is a member of FEBC Radio International. The main transmitting station of FEBA is at Mali in the Seychelles, a Harris 100 kW SW PDM transmitter. FEBA, in common with other religious broadcasting companies, claims to be entirely dependent on financial contributions from Christians.

FEBA broadcasts in the following languages and dialects: Amharic, Arabic, Bengali, Dari, English, Farsi, French, Hindi, Kannada, Malagasy, Malayalam, Marathi, Nepali, Oromo, Portuguese, Punjabi, Pushtu, Sindhi, Sinhalese, Somali, Swahili, Tamil, Telugu, Tigrinya and Urdu.

One African religious broadcaster is Radio ELWA—Eternal Love Winning Africa, located in Liberia. On the face of it, Radio ELWA would appear to have no obvious Western connections; however, the US Information Agency has one of its SW stations at the same location, Monrovia, Liberia. Radio ELWA has no transmitters with more than 50 kW power, but nevertheless its output of programmes in Arabic (15 hours per week) makes it an important player in its region of Africa.

31.1 Renaissance of private US super power

By the late 1970s the ongoing electronics revolution had brought about the cheap, compact, battery powered radio receiver. Millions of such receivers, produced mainly in the Far East, began to flood into America, Europe and particularly the Middle East, partly because of its proximity to Asia. The mass marketing of such receivers with the ability to tune into the short waves did not go unnoticed by the world's broadcasting agencies and religious organisations.

In 1973 the FCC brought in new regulations which permitted the licensing of privately owned radio stations for operation on the SW band in the USA. The first religious station to take advantage of this new legislation was KTRW in Guam.

From then on, growth was rapid, not only in numbers, but also in transmitter output power. From a power of 100 kW some religious broadcasters quickly moved to super power, 500 kW, including KUSW broadcasting from Salt Lake City, Utah, and WCSN Christian Science Monitor broadcasting from Boston. KYOI in Saipan is part of this same network, as is WSHB, which is the most powerful SW station in the private sector, consisting of two 500 kW transmitters coupled in a phased arrangement. With 1000 kW SW output, this station qualifies as probably the largest in the world, setting a new standard in short wave super power.

All of these broadcasting stations are propaganda directed in the broadest sense, carrying commercial advertising, religion or political views. The most effective method is a combination of all three. In terms of target areas, some stations concentrate on the Far East, in particular China; Taiwan (the 'Republic of China') relays programmes from Florida via WYFR for rebroadcasting to mainland China. Collectively, these private radio stations cover every zone of the world and over 80 languages. There is an element of co-operation rather than competition, not normally associated with private enterprise. In 1989 the number of privately licensed SW broadcasting agencies in the USA stood at 19, with another six said to be 'in the pipeline' (Table 31.1).

Table 31.1 *Privately licensed international broadcasting stations*

Call sign	Location	Quantity	Power, kW	Licensee
WYFR	Okeechobee, Fl.	11	100	Family Radio
WYFR	Okeechobee, Fl.	2	50	Family Radio
KGEI	Redwood City, Ca.	1	250	Voice of Friendship
KGEI	Redwood City, Ca.	1	50	Voice of Friendship
W1NB	Red Lyon, Pa.	1	50	World Int. Broadcasters
KTRW	Guam	4	100	Trans World Radio, Pacific
WRNO	New Orleans, La.	1	100	Worldwide WRNO
KHBI	Saipan	2	100	Christian Science Monitor
KFBS	Saipan	3	100	Christian Science Monitor
KSDA	Guam	2	100	Adventist World Radio
WHRI	Noblesville, In.	2	100	World Harvest Radio
KVOH	Ranch Simi, Ca.	1	50	High Adventure Ministries
KCBI	Dallas, Tx.	1	50	KCBI International
KNLS	Anchor Point, Ak.	1	100	World Christian Broadcasting
WMLK	Bethel, Pa.	1	50	Assemblies of Yahweh
WCSN	Scotts Corner, Me.	1	500	Herald Broadcasting
KUSW	Salt Lake City, Ut.	1	500	KUSW
WSHB	Cypress Creek, Sc.	2	500	Herald Broadcasting
WWCR	Nashville, Tn.	1	100	Worldwide Christian Radio

31.2 Religion and politics

All religions have strong links with politics and wars, and Christianity is no exception. Christianity has featured in countless wars over the centuries. During the Second World War both sides prayed to the same God for victory. So strong was the faith in God that even in the last few days of the war, when defeat stared the Germans in the face, Dr Goebbels sustained resistance among the German people with his passionate belief that divine intervention would save the nation from the jaws of defeat.

After 1945, the USA became the first nation to enlist the aid of religious broadcasting in its fight against communism. Religion was (accurately) perceived as a unifying force. Religion deals in clear-cut issues, which people can easily understand: a moral conflict between the forces of Good, identified as Christ, and Evil. From this, it is a simple extension to identify the forces of tyranny and evil with Evil and the Anti-Christ.

The use of radio broadcasting as a means of extending the Gospel has long been common in America. The Association of Religious Broadcasters was formed in the USA 47 years ago, and proved to be one of the most successful ways of raising funds. Today, religious broadcasting in America is big business, creating wealth and power. The most powerful television station in America is Channel 55, the so-called Super Channel, projecting the word of God to the Gulf of Mexico.

All religious broadcasting in the USA is done through privately owned and registered broadcasting stations. It is eay to see how the possession of a powerful voice on the national air waves can raise revenue. Less easy to rationalise is how the international religious broadcasting agencies can pay their way, since these radio stations are often broadcasting to the poor countries of the world.

Paradoxically, however, international religious broadcasting is one of America's growing industries. Each radio station claims that it is funded entirely from voluntary contributions, yet some of these SW broadcasting stations are investing in the latest and most powerful broadcasting technology with the object of projecting a powerful voice into alien countries. The capital costs of modern transmission technology and the super transmitters themselves, added to the cost of installation, staffing and the energy costs of operating a megawatt-powered radio station, raises legitimate questions as to the real purpose behind these broadcasting installations and the source of their financial backing.

There may be a parallel between the early history of Radio Free Europe/Radio Liberty and international religious broadcasting. RFE/RL were set up in the 1950s, and claimed to be privately owned radio stations whose operating costs were funded by charitable contributions. Several years later, under the US Freedom of Information Act, it was revealed that these radio stations in Germany were funded by the CIA, beyond the scrutiny of Congress and with their budgets concealed from Congress and taxpayers alike.

From the time the Freedom of Information Act was passed, the operations of RFE/RL became common knowledge, and RFE/RL ceased to have any covert usefulness in propaganda broadcasting. President Nixon was moved to comment on the need to retain covert intelligence operations. 'What we can do is to give covert support to indigenous political organisations that support our position . . . Unless we restore the covert capabilities of CIA we are going to get rolled by the Soviets.'

One could speculate that the vacuum left by the shift in character of RFE/RL to an overt broadcasting agency such as VOA has been filled with the religious broadcasting agencies. It is a fact that the past decade has seen the greatest ever growth in international religious broadcasting. Within the past five years, three 500 kW transmitters have come into service. In February 1990 the religious broadcasters held their 47th convention in Washington, DC; the convention was opened by President Bush, perhaps the best indication of its importance in world affairs. Those who attended the convention included the manufacturers of super power transmitters and other suppliers of technical equipment.

The advent of modern delivery systems such as microwave and satellite links has had a considerable impact on religious broadcasting. It has made it possible for one programme producing agency to feed hundreds of radio stations, AM and FM, SW and MW with the same programmes. Another growth industry is in the making of recorded programmes for rebroadcasting over religious networks. The contents of some of these tapes seem to have little to do with religion, pointing unmistakably towards their political motivation.

31.3 FCC licensing requirements

In the USA, privately owned SW broadcast stations are licensed by the Federal Communications Commission in accordance with Part 73, Subpart F of its Rules & Regulations. These define an international broadcasting station as: 'A broadcast station employing frequencies allocated to the broadcast service between 5950 and 26100 kHz, the transmissions of which are intended to be received directly by the general public in foreign countries. (A broadcast station may be authorised more than one transmitter.)'

The applicant must satisfy the FCC that there is a 'need' for the international service, that the 'public interest, convenience, and necessity will be served' through the operation of the station. FCC rules also specify that: 'A licensee of an international broadcast station shall render only an international broadcast service which will reflect the culture of this country and which will promote international goodwill, understanding and co-operation. Any programme solely intended for, and directed to the US does not meet the requirements of this service.'

International broadcasts can include commercial or sponsored programmes, provided that the commodity advertised is regularly sold or is being promoted for sale on the open market to which the programme is directed. These FCC rules do not apply to US Government broadcasting agencies, although even these are precluded by the terms of operation from operating within the USA.

One commentator, Hinton, in an article published in *Religious Broadcasting* in February 1990, began by saying: 'Who could have predicted the drastic changes that have occurred this year in Eastern Europe, who knows what potential lies ahead in the Soviet Union, where the distribution of bibles and freedom of preaching is now surpassing all expectations? And who knows what may happen in China if we keep the airwaves beaming the messages of Christ throughout the Orient?' Hinton summarises in a few sentences the potential of religious broadcasting. The Baltic States have been the subject of regular nightly broadcasts by VOA, Radio Liberty and others for a number of years. VOA and RL both broadcast in Estonian, Latvian and Lithuanian on as many as 13 frequencies at different times.

One of the most powerful privately owned broadcasting agencies in the USA is WYFR Family Radio. From its eleven 100 kW and two 50 kW transmitters in Okeechobee, Florida, it broadcasts in many different languages, including Russian. Its motto is 'That all the peoples of the earth may know that the Lord is God and that there is none else.' WYFR has a relay agreement with Taiwan, whereby its programmes are rebroadcast from Taiwan and directed across the straits to the People's Republic of China from as many as six transmitters. This relay service calls itself the Voice of Free China (VOFC). As with other broadcasters in the USA, WYFR is a private company—but it may be stretching credibility to believe that it is not sponsored with a public service grant through some agency.

Relative to its geographical size (about the same as Wales), Taiwan has more broadcasting power than any other country in the world, much of which is directed across the narrow straits towards its political enemies in the People's Republic of China. The Broadcasting Corporation of China (BCC) has some 30 stations, and there is also the Central Broadcasting System (CBS) and VOFC. Much of this output is beamed to the People's Republic of China. Propaganda broadcasting may or may not have contributed to the unrest that triggered the Tiananmen Square massacre; certainly, there was considerable broadcasting activity before the uprising took place.

There is every sign that religious broadcasting from privately owned companies in the USA over the next two decades will grow at an accelerating rate, with 500 kW super transmitters becoming more popular. Religious broadcasting is not seen as a replacement for the more traditional propaganda approach typified by VOA, but operates in parallel with conventional broadcasting, which has taken on a more overt status as an instrument for the projection of a culture, rather than simply of political beliefs.

31.4 World's biggest SW station

The world's most powerful, privately owned SW transmitter, completed as part of a turnkey project in Cypress Creek, South Carolina, USA, was handed over to World Service Herald Broadcasting WSHB on 1 April 1989. The station is one of a small handful of SW stations with 1000 kW of carrier power in the world.

The station was ordered by the First Church of Christian Sciences in Boston, Massachusetts, for Herald Broadcasting, a division of the Church's subsidiary, Christian Science Monitor Syndicate. This new transmitting station, one of the most modern in the world, and certainly the most powerful religious station in the world, will broadcast news and religious programmes in English and Spanish. Programmes will cover the whole of Latin America.

Cypress Creek transmitting station consists of two 500 kW super power SW transmitters of ABB manufacture, the outputs of which are fed into phased high-gain SW curtain antennas. The transmitters are the latest generation of frequency-agile fast-tuning high-efficiency pulse-step-modulation design. Additionally the transmitters incorporate dynamic carrier control, which gives even greater power saving over existing designs of transmitters.

Figure 31.1 *500 kW ABB SW transmitter. On the right the pulse-step-modulation and the transmitter control system, on the left the high frequency stages and in the centre the high power final stage transmitter valve*

It is appropriate, in a way, that the religious organisation Christian Science Monitor should break new ground in the world of private SW broadcasting, because it also has the distinction of being the first licensed religious broadcaster in the US—it was first licensed in early 1930. Religious broadcasting has now become the fastest growing sector in international SW broadcasting. Growth has been particularly spectacular since the early 1980s, and it is not restricted to the USA; there are many new stations in the Middle East and the Far East. Moreover, there is a definite trend towards the use of super power. There are now four such stations in the USA, and more are being planned.

Chapter 32

Transmitter sales during the 1980s

From 1945 and well into the 1960s, it was the US manufacturers that dominated the market for transmitters for international broadcasting. There were a number of reasons for this. During the war, the USA was actually building up an electronics industry in preparation for post-war domination of Europe. In a war-torn Europe the manufacturing industries of Germany, France and the Netherlands had almost ceased to exist in any competitive shape, and although Britain's manufacturing industries had not suffered the same devastation as other European countries it had lost its export markets because its industries had all been turned over to military production during the war.

Contributing to US domination during the post-war years was the Marshall Plan, which had the effect of making Europe even more dependent on American manufactured goods. In the case of broadcast transmitters, the post-war build up of propaganda broadcasting capability by VOA, the BBC and Britain's other broadcasting agency, which was operated by the Diplomatic Wireless Service, was powered by US manufacturers. The two US companies that dominated the transmitter market in America and in parts of Europe were Gates (which later became Harris) and Continental Electronics. To their credit, both companies succeeded in producing some fine examples of AM transmitter designs. The 100 kW transmitter designed by Harris became popular with many broadcasters, and Continental built up a range of AM transmitters that extended from 50 to 2000 kW.

Surprisingly, Continental achieved success at both ends of this power range and, indeed, for its entire range in between. 50 kW is now considered as low power in MW international broadcasting, yet in the first two decades after 1945 it sold over 150 of this model, and the figure today is likely to be more like 200 units. One model was sold to such respected broadcasting agencies as the BBC, Diplomatic Wireless Service, VOA, Trans-World Radio, Voice of Free China and many others.

At the top end of the power range Continental enjoyed much success with its type 320-C, a 1000 kW LW/MW transmitter. In a dualled arrangement it sold five 2000 kW transmitter systems to Saudi Arabia and one each to Egypt, Jordan and Yugoslavia. In addition, nine 1000 kW transmitters were sold to various countries, four for VOA and installed at Germany, Okinawa, Philippines and Thailand.

From the 1970s European companies were in the ascendant: ABB, AEG, Marconi, Siemens and Thomson. These companies had been investing heavily in transmitter development since the 1960s; most worthwhile achievements in transmitter design have been the result of heavy R&D spending in transmitters and related components. ABB, Siemens and Thomson-CSF were also responsible for the progress that had been made in high-power transmitting tubes.

In the six years after the end of the Second World War, Thomson breached the 100 kW barrier in SW transmitter design with its 250 kW transmitter, a noteworthy milestone in transmitter technology. In 1972 the same company raised the power stakes to 500 kW in SW transmitters, a power output achieved with a single tube in the final RF stage. AEG and ABB brought out similar transmitters at about the same time. In 1982 AEG achieved another major advance when it brought out its Pantel range of AM transmitters equipped with pulse-duration-modulation, which enabled transmitters to operate at much higher efficiencies. Five years later ABB introduced its PSM range of transmitters, capable of even greater operating efficiencies.

32.1 LW/MW sales

Growth in activity on the MW and LW bands has almost equalled that for the short waves, and as with the short waves it has been of a multidimensional character. Growth has taken place in numbers of countries, numbers of broadcasting agencies per country, numbers of transmitting stations per broadcaster and in the average output power for a transmitter.

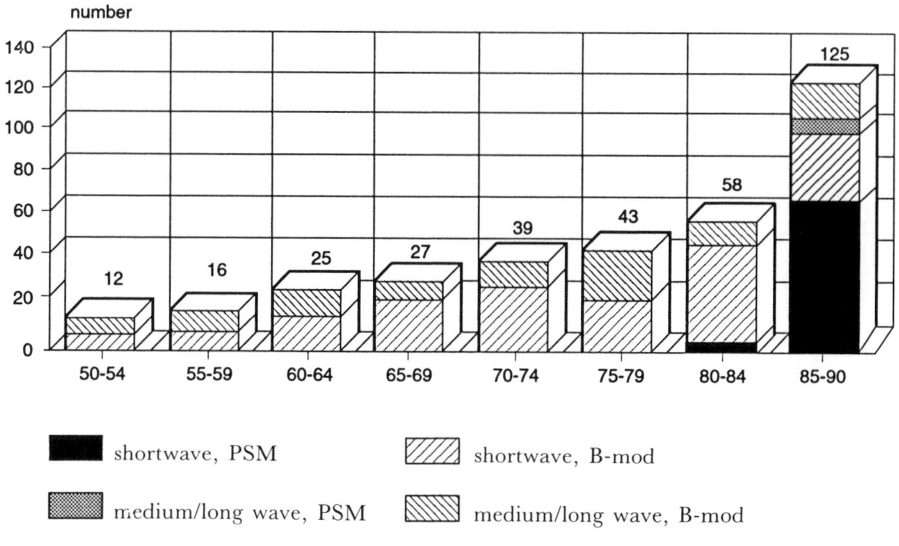

Figure 32.1 *Total sales of ABB transmitters*

To carry out a complete survey would require much research, involving polling each country and its broadcasters. An easier approach is to analyse sales figures over the past 50 years. The difficulty with this method is distinguishing growth from equipment replacement. However, in the case of broadcast transmitters the problem is greatly eased by long average life of a transmitter. The BBC is not alone in having kept MW transmitters in service for 50 years or more. Transmitters for broadcasting on the LW and MW bands tend to have a longer life in service than those for broadcasting on short waves.

There are two reasons for this. The technology of SW transmitters has made more advances, particularly in high-speed tuning, high efficiency operation and unattended operation, whereas in the technology for MW/LW it has advanced in efficiency and power output. Thus, if we take the sales figures for MW/LW transmitters over the past 50 years it is reasonable to assume that those less than 30 years of age, and perhaps 35% of those older than 30 years, will be in service.

There are more than 20 different manufacturers in the world of AM transmitters for MW/LW broadcasting, but most of these restrict their range to low- to medium-power range. At the high-power end, there are only five players, four of them in Europe: ABB, Radio Industries Zagreb, Tesla, and Thomson-CSF, plus Continental Electronics in the USA.

32.2 Thomson-CSF

Of these companies, the leaders in sales of super transmitters are ABB, Thomson-CSF and Varian-Continental. By my estimates, Continental has sold 25 1000 kW units, ABB 21, RIZ seven and Thomson-CSF 20. Some of these units were in a dualled arrangement to give 2000 kW carrier power, and again the leading company on past sales is Continental with eight 2000 kW transmitters. However, in the past ten years Thomson-CSF and ABB have outstripped all comers. Thomson-CSF moved into top place when it sold three 2000 kW broadcasting stations in 1985 to a Middle East state.

Sales in the 1950s represented mainly re-equipping of national broadcasting networks, with no major advances in transmitter technology. The 1960s saw advances in higher output powers made possible through developments in tube designs for high-power transmitters. The 1970s were characterised by increased

Table 32.1 *Thomson-CSF transmitter sales*

Period	Units	Total kW
1950-54	10	860
1955-59	5	700
1960-64	12	2100
1965-69	22	7800
1970-74	40	19 275
1975-79	22	10 200
1980-84	26	12 900
1985-89	17	9600

Figure 32.2 *TH 558 Tetrode tube for 500 kW SW and 600 kW LW/MW service*

sales and an increasing trend towards super-power broadcasting. This period coincides with a shift in world wealth in favour of the Middle East, particularly OPEC member states. Some of the income from oil revenues was pumped into the acquisition of super-power broadcasting capability.

In the 1980s, the total sales were roughly the same as for the previous decade. Since none of these sales are replacement, the growth rate works out roughly at doubling the transmitter power in the world.

Thomson-CSF's sales from 1950 total 154 transmitter units and 63 435 kW. Continental's sales over the same period total 218 units and 42 000 kW. RIZ has sold some 68 units totalling 13 300 kW, AEG-Telefunken some 81 units and 23 400 kW (AEG's range of MW transmitters does not exceed 600 kW). To a limited extent, RIZ's transmitter sales are also confined to carrier powers below 600 kW.

Thomson-CSF has a number of companies and divisions. The following sales relate to the AM Broadcast Department at Gennevilliers.

Table 32.2 *Thomson-CSF high-power LW/MW transmitter sales*

Date	Country	Location	Transmitter	
1967	Romania	Brasov	1200 kW (2×600)	LW
1968	Saudi Arabia	Riyadh	1200 kW (2×600)	MW
1972	Turkey	Ankara Polatu	1200 kW (2×600)	LW
1972	Luxembourg	Beidweiller	2000 kW (2×1000)	LW
1972	Iraq	Basrah	1200 kW (2×600)	MW
1973	Germany	Felsberg Eur 1	1000 kW	LW
1973	Germany	Heusweiller	1200 kW (2×600)	MW
1974	France	Allouis	1000 kW	LW
1974	Monaco	Roumoules	2000 kW (2×1000)	LW
1974	Turkey	Istanbul	1200 kW (2×600)	MW
1977	Iran	Abadan	1200 kW (2×600)	MW
1978	Germany	Felsberg Eur 1	1000 kW	LW
1981	France	Allouis	1000 kW	LW
1981	Germany	Felsberg Eur 1	1000 kW	LW
1983	Iran	Chabahar	1200 kW (2×600)	MW
1984	Iraq	Tanaf	2000 kW (2×1000)	MW
1985	Iraq	Maisan	2000 kW (2×1000)	MW
1985	Iraq	Sulaimaniya	2000 kW (2×1000)	MW
1982	Morocco	Nador	2000 kW (2×1000)	LW
1983	Monaco	Roumoules	1000 kW	LW
1987	Monaco	Roumoules	1000 kW	MW
1989	Turkey	Catalca	1200 kW (2×600)	MW

Source: Thomson-CSF customer list edition March 1989.

32.3 Radio Industries Zagreb

Along with Tesla in Czechoslovakia, RIZ is one of the largest transmitter manufacturers in Eastern Europe. From 1951 to 1988, RIZ manufactured over 600 radio transmitters of various types and power rating from 20–2000 kW.

Table 32.3 relates specifically to AM, MW and LW with powers from 1000 kW upwards. It needs to be remembered that transmitters with such high output powers did not come into being until the late 1950s and then only in small numbers.

Table 32.3 *RIZ sales*

Year	Country	Transmitter	Quantity
1968	India	MW 1000 kW (2×500)	2
1974	Libya	MW 1000 kW	1
1978	Libya	MW 1200 kW (2×600)	2
1978	Libya	MW 2000 kW (2×1000)	2
1979	Syria	MW 1000 kW	1
1981	Yugoslavia	MW 1200 kW (2×600)	1

Figure 32.3 *TH 539, the world's most powerful transmitting tube with a rated carrier power output of 1.5 MW. Weight: 155 kg, height: 885 mm, width: 410 mm*

32.4 Continental Electronics

Continental Electronics built the first 1000 kW LW transmitter in the world, which went into service with VOA, installed in Munich for propaganda broadcasting. This first attempt at breaking the megawatt barrier was achieved by combining two 500 kW units. Subsequently the company went on to manufacturing single transmitters with 1000 kW rating.

32.5 SW transmitters

For all the quite remarkable growth that has taken place in MW/LW transmitters, the greatest growth has taken place in SW broadcasting. MW/LW high power

Table 32.4 *Continental Electronics sales of 1000 kW+ transmitters*

Country	Location	Transmitter
Germany	Munich (VOA)	1000 kW LW
Okinawa	Tyukyu Island (VOA)	1000 kW MW
Philippines	Poro (VOA)	1000 kW MW
Thailand	Ban Phachi (VOA)	1000 kW MW
Egypt	Alexandria (OBTF)	1000 kW MW
Venezuela	Coro	1000 kW MW
Taiwan	Taipei (BBC)	1000 kW MW
Saudi Arabia	Qurayat (MOI)	1000 kW MW
Saudi Arabia	Jeddah (MOI)	1000 kW MW
Yugoslavia	Belgrade	2000 kW MW
Saudi Arabia	Duba (MOI)	2000 kW MW
Saudi Arabia	Qurayat (MOI)	2000 kW MW
Saudi Arabia	Duba (MOI)	2000 kW MW
Saudi Arabia	Jeddah (MOI)	2000 kW MW
Jordan	Ajlun (HBS)	2000 kW MW
Saudi Arabia	Damman (MOI)	2000 kW MW
Egypt	Batra (OBTF)	2000 kW MW

Source: Continental customer list, September 1986

Figure 32.4 *RIZ 1200 kW MW broadcasting station in Zadar, Croatia*

transmitter sales peaked in the 1970s, for two reasons: European broadcasters moved towards the use of higher powers on these wavebands, with the object of projecting stable broadcasts over large sectors of mainland Europe and North Africa, and VOA became interested for different purposes. Although growth in MW/LW broadcasting will continue, the spectrum is limited and can support only so many super transmitters.

SW broadcasting has seen much cyclic, multidimensional growth taking place over the past four decades. All this is reflected in Table 32.5 on SW sales. There are many transmitter manufacturers around the world, but as it was necesssary to have a cutoff point, the data has been restricted in the main to the world's major companies, specifically those who manufacture 500 kW SW transmitters.

On top of the figures given, Continental Electronics has sold over 100 transmitters, totalling over 18 000 kW, and Marconi Electronics Inc., USA, has sold 15 transmitters, totalling 7500 kW.

Table 32.5 *Sales of SW transmitters since 1950*

Period	ABB		Thomson-CSF		Marconi		AEG	
	Units	kW	Units	kW	Units	kW	Units	kW
1950-1954	4	400	1	250	4	400	1	50
1955-1959	7	700	2	600	3	300	0	0
1960-1964	13	2350	14	1530	27	4350	22	1950
1965-1969	18	4350	10	1350	26	6050	13	1300
1970-1974	29	7350	25	8150	11	1850	7	1450
1975-1979	19	6450	17	5350	11	3050	14	6750
1980-1984	47	14 750	27	8950	15	5300	16	6750
1985-1990	106	30 500	32	12 600	21	6750	26	10 200
Total	243	66 850	127	37 780	118	28 050	99	28 450

Table 32.6 *AEG Telefunken sales of 500 kW SW transmitters*

Year	Country	Location	Quantity
1982-84	Norway	NTA Kvitsoy	2
1982	Germany	BR Ismaning	1
1984	Vatican	SM di Galaria	1
1982-83	Austria	ORF Moosbrunn	1
1983-84	Netherlands	PTT Flevoland	4
1982-85	UK	BBC Rampisham	4
1986	USA	VOA Greenville	1
1986	Norway	NTA Sveito	1
1987	Japan	KDD Yamata	2
1987-89	Germany	DW Wertachtal	6
1987	Austria	ORF Moosbrunn	1
1989	Portugal	RDP S Gabriel	1
1990-93	Iran	IRIB Sirjan	10

Source: AEG Telefunken Customer List: private communication, 28 March 1990

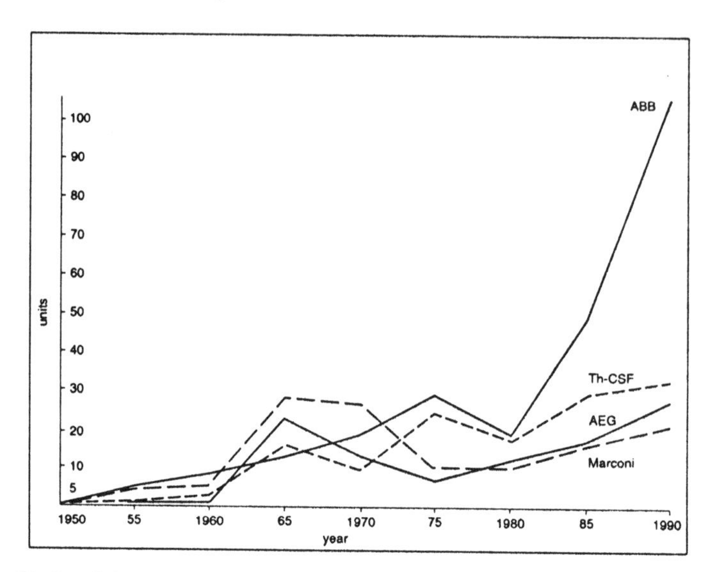

Figure 32.5 *Sales of shortwave transmitters by the big four transmitter manufacturers since 1950*

Sales figures apply to transmitters with powers 100–500 kW from 1960 onwards. For the period before some sales of 50 kW have been included. Note: Sales figures are not cumulative but relate to a five year period. For example, ABB sold 47 units between 1980 and 1985

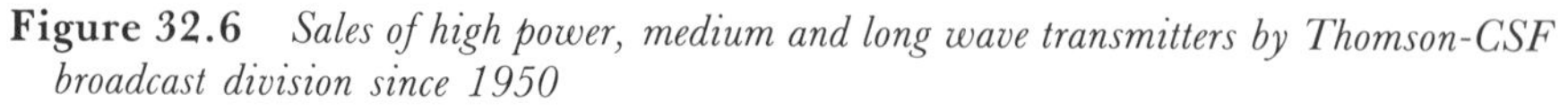

Figure 32.6 *Sales of high power, medium and long wave transmitters by Thomson-CSF broadcast division since 1950*

Note: Sales figures are not cumulative but relate to five year periods. kW sold = number of transmitters × power ratings

Table 32.7 *ABB sales of 500 kW SW transmitters since 1985*

Year	Country	Location	Quantity
1986	USA	VOA Greenville	1
1987	Finland	Yleisradio Pori	1
1987	Yugoslavia	Bijelina	4
1988	Vatican Radio	Santa Maria di Galeria	2
1988	Finland	Yleisradio Pori	2
1989	Jordan	Qasr Kherane	3
1989	India	AIR Bangalore	2
1989	USA	Cypress Creek	2
1989	Kuwait	Kabd (MOI)	2
1990	Kuwait	Kabd (MOI)	4
1990	India	AIR Bangalore	4
1990	Switzerland	PTT	1
1992	Turkey	PTT Emirler	5
1992	Kuwait	Kabd (MOI)	4

Source: ABB news releases 1986-1990

Table 32.8 *Thomson-CSF sales of 500 kW SW transmitters*

Year	Country	Location	Quantity
1972	Sweden	Karlsborg	1
1973	France	Issoudun	4
1973	Sweden	Horby	2
1976	Norway	Frederikstad	1
1977	Libya	Tripoli Sabrata	4
1978	Gabon	Moyabi	4
1981	Iraq	Babylon	2
1981	Malaysia	Kajang Penang	2
1982	Syria	Damas	4
1982	Monaco	Fontbonne	1
1984	French Guiana	Montsinery	3
1984	Iraq	Balad	16
1986	Ivory Coast	Abidjan	1
1988	French Guiana	Montsinery	1
1989	Gabon	Moyabi-Bis	1
1990	Saudi Arabia	Riyadh	4

Source: Thomson-CSF customer list, March 1989

Up to 1989 Continental Electronics had sold only one 500 kW SW transmitter, to VOA Greenville, and Harris Corporation has not produced SW transmitters with power outputs exceeding 100 kW. However, this company did produce the world's first PDM transmitter, and has the capability and resources to produce 500 kW versions.

Table 32.9 *Marconi sales of 500 kW SW transmitters*

Year	Country	User/Location	Quantity
1984	UK	BBC Rampisham	4
1985	Dubai	DRC TV	1
1985	USA	VOA	1
1989	USA	VOA	2
1990	USA	VOA	8
1992	USA	VOA	7
1991	UK	BBC Rampisham	2

Source: GEC-Marconi customer list

Table 32.10 *Summarised statistics on sales of 500 kW SW transmitters*

Company	Sales by type	Total sales to 1990
ABB	35 Class B, 46 + PSM	81 +
Thomson-CSF	44 Class B, 7 + PDM	51 +
AEG	14 Class B, 35 + PDM	49 +
Marconi	0 Class B, 25 + PDM	25 +
Continental	1 Class B, 1 PDM	2
Total sales to year 1990		208

These figures have been compiled from manufacturers' reference lists and are reasonably correct at time of writing. As a rough guide to future sales, orders for a further 25 for 1991 is thought to be realistic.

Table 32.11 *Total computed sales of high-power SW transmitters (1990–1993) by the major manufacturers*

Company	kW			Total no. of transmitters	Total kW sold
	100	250/300	500		
ABB	12	22	28	62	20,700
Thomson-CSF	4	2	23	29	12,400
Telefunken	0	0	10	10	5,000
Marconi GEC	0	11	23	34	14,250
Continental	14	0	4	18	3,400

Percentage share of world market by each of the named companies

Asea Brown Boveri	37%
Thomson-CSF	22·2%
Telefunken	6.1%
Marconi GEC	25·5%
Continental	8·9%

Chapter 33

The future of international AM broadcasting

When I first proposed the idea to the publishers of writing an account of the origins and growth of propaganda broadcasting in the HF spectrum, it was in the conviction that AM broadcasting would not only survive to the 21st Century but would become the main gateway for the dissemination of information and propaganda well into the third millenium.

This forecast was partly founded on a knowledge of the politics of broadcasting formed in the past several years, and partly because of an involvement with the technical side of broadcasting over a much longer period. However, this view was not a universally shared opinion; ever since the launching of the first communications satellite, experts have been forecasting the demise of the HF spectrum as a medium for communications and broadcasting.

International politics since 1945 has been eventful, and international broadcasting on the SW bands has played no small part. The Berlin Blockade (1948), the Hungarian Uprising (1956), the invasion of Czechoslovakia by Soviet troops (1968), the birth of Solidarity and introduction of martial law in Poland (1981) have all taken place. Following these events, the USIA began to plan an 'improved audibility programme' for VOA, to be accomplished by building super transmitters. VOA was following a pattern which had already been set by other international broadcasters such as the BBC, Deutsche Welle and RFI.

In the late 1980s the World Administrative Committee for HF Broadcasting recommended the introduction of single-side-band broadcasting in the HF spectrum by not later than 2015. This recommendation was not greeted with enthusiasm by some of the world's major broadcasting authorities, for reasons which have been discussed in this book. Nevertheless, the recommendation was endorsed. The introduction of SSB broadcasting would have the effect of doubling the available number of frequency allocations in the HF spectrum (3–30 MHz), providing scope for expansion in numbers of SW stations by that time—thus it was a fairly safe prediction that information broadcasting in this spectrum had an assured future.

In the mid 1980s emerged a new user of the short waves: religious broadcasting stations. Religion had used the short waves since the 1930s, but these new broadcasters were different, harnessing power transmitters and religion as a vehicle for propaganda broadcasting. With the benefit of hindsight, it is now evident that the harnessing of religion with world politics was destined to become a potent force in broadcasting, causing divisions between ethnic groups. Yet it woud have required a farsighted individual to have predicted how propaganda broadcasting in conjunction with other media (television) would change the whole political structure of Eastern Europe by a process of revolutions. To paraphrase the words of a former ambassador to Berlin, 'No one expected Eastern Europe to topple that easily.'

Those who doubt the importance attached by Western governments to the power and potential of propaganda broadcasting should study the many pronouncements on the subject made by statesmen and broadcasting agencies. A statement issued by the BBC International Press Office on 29 March 1990 contained some comments on the dramatic events in China and Eastern Europe. After the Tiananmen Square massacre, a listener wrote in to say 'You'll go down in history. We'll remember you forever'. With reference to the more recent events in Eastern Europe, the BBC quoted from some listeners' letters. From Romania: 'I cannot find words to thank you for everything you have done for this country since 1944, but especially during the last two months.' From a listener in Czechoslovakia: 'Your support for democracy and freedom of expression has not been in vain.'

In another press statement, 25 April 1990, the BBC said that 'major international developments in the past year had demonstrated in the most vivid way possible, the immediate and long-term impact of BBC World Service broadcasts on their audiences.' In that same month the BBC World Service prepared a written evidence for the Foreign Affairs Committee as part of its submission for additional funding during the forthcoming year. The summary said that 'since the Committee's last hearings on World Service funding a year ago, major international developments in China, Eastern Europe and the Soviet Union have demonstrated in the most vivid possible way the immediate and long term impact of World Service broadcasts on their audiences.'

In Tiananmen Square, the Chinese student supporters of democracy held high a banner saying 'Thank you, BBC'. In Eastern Europe, testimonies from personalities as diverse as President Havel of Czechoslovakia, the Prime Minister of Poland, and Oleg Gordievsky, a former KGB colonel who became a key western defector, have spoken of the vital importance of the World Service broadcasts in shaping political changes in the former Communist Bloc.

The summary went on to say: 'These and many other instances around the world have confirmed our view that the funds devoted to broadcasting in the national interest do yield results, though over a long term and not in a predictable way or place. They encourage us to present our case for better funding of programmes, of research, of publicity and of our recent and increasingly effective local rebroadcasting effort. . . The importance and impact during the past year of several language services such as all the Eastern European services, Russian, and of course Chinese, exceeded all expectations. The need for the BBC radio services "being there" to meet a crisis was demonstrated and underlined. We have included proposals about language services in our ten year strategy.'

The report goes on to talk about its proposals to 'meet the further efffects of the Information Explosion as it was developing in the USSR and Eastern Europe.' It also goes on to recommend urgent attention to the need to improve coverage in the Indian subcontinent, the Middle East and China. It also makes references to the need to improve the broadcasting facilities in the Gulf (Masirah Island) and Thailand.

33.1 World Service in the 1990s

Perhaps the clearest indication of the importance attached to political information broadcasting by the UK Foreign Office is given by the first changes to World Service broadcasting issued by the FO. These changes follow the BBC's submission for funding. (Although the BBC World Service has a certain degree of editorial licence, it is the FO that decides where and when the BBC should broadcast to various countries in the world. Shifting world politics means shifting to different target zones of the world.)

On 29 May 1990 the BBC was instructed to suspend broadcasts to Japan and Malaya, and to use the saved programme time to expand broadcasting in Russian, Mandarin and Vietnamese. Before this announcement the BBC was claiming much success with its broadcasting to Japan, reaching a listener audience of 225 000 for a cost of £279 000. By any standards this was a remarkable performance; many newspapers and satellite companies are willing to sink millions to reach such a sizeable audience.

The FO, however, has never been noted for its display of business acumen. What the FO lacks in business acumen is more than compensated by its passion for world politics. The ideal solution would have been an increase in transmission capacity, but it would take several years to provide this: to plan new transmitting stations like Tsang Tsui, the new SW station with a 500 kW broadcasting capacity. One of the reasons for the new station was claimed to put a stronger signal into Japan, but this was a cloak for its real purpose; to put a stronger signal into the People's Republic of China.

Evidently the FO had a reason for not waiting until another new station could be constructed, and a switch in programmes from one country to another can be done almost immediately, so out goes the Japanese service and China comes in for an even greater share of the total coverage of the BBC World Service. The increase in BBC World Service broadcasts to China occurred a few days before the first anniversary of Tiananmen Square.

33.1.1 Comparisons with BBC national broadcasting

Although this book is not concerned with national broadcasting, it is necessary to make reference to this from time to time. The best indication of where the government and national interest lies is given by a comparison of the two capacities in terms of broadcasting power. As described earlier, the BBC World Service has a SW capacity of 20 000 kW, its MW/LW stations 4200 kW.

For national broadcasting, the BBC radiates five separate programmes Radio 1 to 5. The total capacity of all AM stations with powers in excess of 50 kW comes to under 2000 kW.

One of the disadvantages of MW/LW AM transmission is the high cost of the transmitters, antennas and the estate needed to support the massive tower structures. Accordingly, the BBC has for some time been running a publicity campaign to sell VHF to the listening public, gradually shifting more and more popular services to VHF.

Yet, even while the BBC is telling its British listening public that AM is 'steam radio', and will eventually be phased out, it is investing hundreds of millions of pounds on AM transmitters for the service it puts out to listeners all over the world, mainly in China and Russia. And although British listeners have to buy a licence for the privilege of listening to national broadcasts, overseas listeners get the programmes free of charge.

In terms of capital investment for new transmitters to replace ageing equipment, the comparison is even more marked. Transmitter stations of the World Service get refits every several years, or additions to existing transmitters. For its national broadcasting services the BBC implemented a refit in 1978 of several transmitting stations (Brookmans Park, Moorside Edge and Droitwich), and the transmitters taken out of service were in some cases 50 years of age. This is a tribute, of a sort, to the company that supplied the transmitters of both generations (Marconi), but it is also an indictment of the low priority given to entertainment broadcasting, compared to the broadcasting in the national and government interest on the same AM wavebands.

So long as there is such a difference in priorities, we may be confident about the long-term future of information propaganda broadcasting in the SW, MW and LW bands.

33.2 World Service global audience

The BBC's total audience in all languages breaks down as follows:

Europe	31 million
Africa	20 million
Arab world	10 million
South Asia	53 million
Far East	3 million
Latin America	2 million
Rest of World	1 million

25 million listen to World Service in English:

Europe	3 million
Africa	10 million
Arab world	1 million
South Asia	9 million
Far East	1 million
North America, Canada, Australia	1 million

Regular listeners are defined in international broadcasting terms as adults listening once a week or more. The BBC's 31 million European audience includes 26 million in former Communist countries.

Recent surveys in India in two states reveal 7 million listeners in Uttar Pradesh and 3 million in Maharashtra.

Political crises and major world events can have a major impact on listening. In the first ever survey commissioned by the BBC in Poland, the 1.9 million regular listeners were swollen by an extra 3 million who tune in 'depending on events'.

BBC broadcasts in Pushtu have an audience of 63% of Afghan refugees—higher than any other broadcaster. 25% of Afghan refugees listening in Persian also tune to the BBC, more than any of its competitors.

It is clear that listenership in Eastern Europe is going to change. There are indications of some fall-off of regular listening in Poland and Hungary. An ambitious research programme is being pursued for the first time in Eastern Europe.

In Bangladesh, the BBC is the leading international broadcaster in Bengali (13.4% of the population, about 8.2 million people) and English (more than a million).

Almost one in four adult Nigerians listen regularly to the BBC, in Hausa, English, or both. The BBC audience has increased while numbers listening to other foreign radio stations have fallen. According to a national survey last year, there are 14.7 million regular BBC listeners. A recent survey of 'top people' in Africa confirms the reach of the BBC French Service—in Zaire, one in four are regular listeners. In Tanzania, Zambia and Ghana, the BBC's English broadcasts are clearly in the lead, with 85% of this group regular listeners in Ghana.

South Africa is a difficult challenge. There is widespread access to short wave but South Africans are not used to listening to it. New listeners are being attracted but so far it appears to have been a slow process: 2.4% of urban whites (70 000 people) listen regularly, but only 0.5% of urban blacks (about 35 000).

In Portugal, the decline in audiences has been reversed. About 312 000 adults now listen regularly to BBC Portuguese programmes on short wave, up from 60 000 a few years ago. National rebroadcasts by Radio Renascenca attract about 360 000.

Local rebroadcasts on FM are having a large impact in Finland. In towns where rebroadcasts are available, between six and 13 times as many people are regular listeners to the BBC's Finnish programmes.

Bibliography

1 Books

BAKER, W.J.: 'A history of the Marconi company' (Methuen, London, 1970)
BARNOUW, E.: 'A history of broadcasting in the US'
Volume 1: 'A tower in Babel' (1966)
Voume 2: 'The golden web 1933–53' (1968)
Volume 3: 'The image maker 1953–70' (1970) (OUP, New York)
BOYD, D.A.: 'Broadcasting in the Arab world' (Temple University Press, Philadelphia, 1982)
BRAY, W.J.: 'Memoirs of a telecommunications engineer' (Private publication, 1989)
BRIGGS, A.: 'The BBC—the first 50 years' (OUP, 1985)
BROWN, F.J.: 'Cable & Wireless communications of the world' (Pitman, London, 1929)
DUUS, M.: 'Tokyo Rose—orphan of the Pacific' (Kodansha International, New York, 1979)
ECCLES, W.H.: 'Continuous wave wireless telegraphy' (Wireless Press, 1929)
HALLIDAY, F.: 'The making of the Cold War' (Verso & NLB Publications, London, 1983)
HEAD, S., and STERLING, C.H.: 'Broadcasting in America' (Houghton Mifflin, Palo Alto, 1972)
HENNEY, K.: 'The radio engineering handbook' (McGraw-Hill, New York, 1941)
HERZSTEIN, R.E.: 'The war that Hitler won' (Hamish Hamilton, London, 1979)
HYDE, M.G.: 'Secret intelligence agent' (Constable, London, 1982)
LAWRENSON, J., and BARBER, L.: 'Price of truth—the story of Reuter' (Mainstream Publishing, Edinburgh, 1985)
MOSELY, L.: 'The Dulles family' (Hodder & Stoughton, 1978)
NICHOLS, R.: 'Radio Luxembourg' (W.H. Allen, London, 1983)
NIXON, R.: 'The real war' (Sedgwick & Jackson, London, 1980)
PAWLEY, E.: 'BBC engineering 1922-1972' (BBC Publications, 1972)
RODRIGO, R.: 'Berlin airlift' (Cassel, London, 1960)
SHIRER, W.: 'A history of Nazi Germany—the rise and fall of the Third Reich' (Secker & Warburg, 1960)
SHIRER, W.: 'Mid century journey' (New American Library, 1961)
SPERBER, A.M.: 'Murrow—his life and times' (Michael Joseph, 1987)
STOKES, J.W.: '70 years of radio tubes and valves' (Vestal Press, New York, 1982)
WEST, N.: 'MI6—the British Secret Service' (Weidenfeld & Nicolson, 1983)
WEST, N.: 'GCHQ—the secret wireless war' (Weidenfeld & Nicolson, 1986)
WOOD, J.: 'Satellite communications and DBS systems' (Butterworth Heinemann, 1992)

2 Specialist publications, yearbooks and technical articles

BBC External Services Publicity Unit:
'Voice for the world 1932–1982' (1982)
'1987 Year Book' (1988)
'BBC World Service—voice for the world' (1988)
'Soviet jamming of international radio broadcasts' (USIA, 1987)
'Trans World Radio' (TWR, Chatham, New Jersey, 1989)
'Finnish broadcasting—Pori SW station' (FBC, Helsinki, 1989)
'World radio & TV handbook 1987' (Billboard Publications, Denmark and New York)
'World radio & TV handbook 1989'
'Rugby radio station' (Post Office Telecommunications, 1977)
'Transmision 1989-90 year book' (Thomson Business Publications, 1990)
'The radio amateur's handbook' (American Radio Relay League, 1956)
'History of RIAS' (Special publication marking the 750th anniversary of Berlin, Rundfunk im Amerikanischen Sektor, Berlin, 1987)

3 BBC news releases during 1990

'BBC World Service gives evidence to MPs' (25 April)
'Changes in BBC World Service hours and services' (30 May)
'BBC World Service consolidates its audience lead' (6 June)
'MPs urge inflation protection for World Service' (21 June)
'World Service's trusted role in disseminating truth' (26 July)
'BBC World Service steps up Arabic broadcasts' (2 August)
'BBC confirms Iraqi jamming' (7 August)
'BBC World Service launches new crisis "Gulf Link" ' (5 September)
'Extra funds for BBC World Service' (9 November)
'BBC World Service orders £3.2 million multi-language computer' (14 November)
'BBC World Service boosting output for South Africa' (21 November)

4 Previous articles and papers by the author

WOOD, J.: 'Developments in design and performance of HF transmitters', *Electron. & Power*, June 1987, **33**
'High power transmitters' *Int. Broadcast.*, June 1987, **10**
'Broadcasting the Voice of America', *ibid.*, July/Aug. 1987
'Single sideband AM broadcasting', *ibid.*, Sept. 1987
'Transmitter powers', *ibid.*, Sept. 1987
'Energy saving for AM transmitters', *ibid.*, Oct. 1987
'Transmitters: state of the art Europe', *ibid.*, Sept. 1988, **11**
'The super contract for VOA super transmitters', *ibid.*, Jan./Feb. 1989, **12**
'The world of antennas', *ibid.*, Mar. 1989
'Upsurge in shortwave transmitter sales', *ibid.*, Sept. 1989
'Focus on Europe: Siemens', *ibid.*, Oct. 1989

'Focus on Europe: RIAS Berlin, *ibid.*, Nov. 1989
'Focus on Europe: Asea Brown Boveri, *ibid.*, Dec. 1989
'Focus on Europe: Finnish Broadcasting', *ibid.*, Jan./Feb. 1990, **13**
'Thomson-CSF Broadcast Division', *ibid.*, May 1990
'Radio Monte Carlo', *ibid.*, June 1990
'AM terrestrial broadcasting—past, present and future', *IEE Review*, March 1989, **35**
'AM terrestrial broadcasting—the short-wave super transmitter', *IEE Review*, April 1989, **35**
'Broadcasting with super power from the Arab world, part 1', *Int. Broadcast.*, July/Aug. 1990, **13**
'Broadcasting with super power from the Arab world, part 2', *ibid.*, Sept. 1990, **13**
'Speaking unto nations—international broadcasting with megawatt power', *IEE Review*, June 1990, **36**
'Growth explosion in information broadcasting in the HF spectrum', *Telecomm. Policy*, Feb. 1991
'Desert sounds; International broadcasting from the Arab world', *IEE Review*, 1991, **37**, (7)
'High power from a low profile', *IEE Review*, 1992, **38**, (1)
'New concept targets audiences for Radio France International', *World Broadcast News*, 1992, **15**, (3)
'New wave at RFI', *Int. Broadcast., 1992,* **15**, (1)

5 Manufacturers' reference lists for transmitter sales

Broadcast transmitter customer sales list, Continental Electronics, Dallas, Texas, 1986
Broadcasting transmitters lists of references, AEG Telefunken Berlin, Germany, 1990
List of shortwave transmitter sales and MW 500 kW upwards, Marconi Company, 1989, Chelmsford, UK
Shortwave and medium wave transmitters customer reference list, Thomson—CSF Broadcast Division, France, 1990
Transmitter reference lists, Radio Industries Zagreb, RIZ, Yugoslavia, 1988
Transmitter sales data, Asea Brown Boveri, ABB, Baden, Switzerland

6 Other references

CHESNOY, J.L.: 'Totalitarianism in the 1930s through three newspapers: The Times, The Daily Mail and The Daily Herald' (Universite de Tours, France, December 1983)
SCHMINKE, W.: 'High power pulse-step modulator for short and medium wave transmitters' (Brown Boveri Review, 1985, **72**)
TSCHOL, W., and BOKSBERGER, H.U.: 'Latest developments in high power transmitter sector' (Brown Boveri Review, 1987, **74**)
JAUSSI, A.W., and TSCHOL, W.: 'Measurements of efficiency of high power transmitters with regular broadcast programmes' (EBU Review, 1982, 194)

THOMANN, P., and SCHWARZ, A.: 'Modification of the MW station at Heusweller to DCC' (Rundfunktechn. Mitt., 1988)
SOWERS, M.W., HAND, G., and RUSH, C.M.: 'Jamming in the broadcast bands of the high frequency spectrum', *IEE Trans.*, June 1988, **34**, 2
JACOBS, G.: 'The renaissance of private SW broadcast stations in the US', *ibid.*
WELDON, J.O.: 'The early history of US international broadcasting from the start of World War 2', *ibid.*
BERMAN, G.A., and GARLINGTON, T.R.: 'Evaluation of 500 kW SW transmitters at the Voice of America', *ibid.*
SHAUGNESSY, J.: 'The Rugby radio station', *J. IEE.*, 1926, **64**
WHITAKER, J.: 'Advanced technologies—satellites and computers have dramatically changed communications and broadcasting', *Broadcast Eng.*, 1989, **31**, pp. 74–86
RUSSELL, M.: 'The 420B 500 kW SW transmitter', *Int. Broadcast. Eng.*, 1988)
'Largest ever broadcasting contract' (GEC Marconi news release, 18 July 1988)
'VOA contract awarded to Marconi Electronics Inc./Cincinnati Electronics' (USIA, Washington, news release, 31 May 1988)
TASS news report from Minsk, 26 May 1989
'Contract award from RFE Munich' (ABB news release, September 1989)
'Contract award from RFE Munich' (Thomson-CSF news release, 8 February 1989)
'External broadcasting—estimated programme hours per week of some external broadcasters' (International Broadcasting audience research, BBC, August 1986)
'The report of the President's Task Force on US Government International Broadcasting', prepared in December 1991, Washington DC

Appendix I

Television, the Gulf War and the future of propaganda

The Gulf War was one of the shortest wars in history—but the war, and the run-up to it, gave ample proof that propaganda broadcasting is still capable of living up to the roles it played in the Second World War and the subsequent Cold War.

The Gulf War saw a honing and combining of propaganda techniques developed by the major participants in those wars. Information, misinformation, 'mirror' broadcasting, psychological warfare and covert broadcasting were all present in unprecedented intensity and ferocity. Various forms were used to inform and deceive, justify and confuse, disorient and bring about internal uprisings, sabotage and terror in the enemy country.

The war also saw an important new dimension in propaganda: television. Electronic news gathering and the portable satellite terminals showed television viewers the world over selected scenes of war. Emotive visual imagery was harnessed to the carefully scripted spoken word, skilfully crafted to wring emotion from the viewer.

An interesting feature of television propaganda in the Gulf War was a reversal of the usual biases of oral communication in Arab and Western cultures. Traditionally, Western speakers base their arguments on reason, whereas Arabs use emotive rhetoric. In the Gulf War, Western politicians relied heavily on generating emotion, with bellicose language and invective lacing of every sentence. It was common to talk of the enemy leader as a 'monster' who would be 'brought to justice in chains'. By contrast, Saddam Hussein— when he was not promising 'the mother of all battles' and 'wading in seas of blood'—appeared on Iraqi television showing no emotion, speaking clearly and quietly.

Western leaders use television as a rabble-rousing medium. When President Bush proclaimed on TV 'Here is my message to Saddam: get out of Kuwait!', he was not addressing the Iraqis. When politicians wish to mediate they use diplomatic channels, secure and private; when they wish to confront, they use open forms of mass communication.

The role of television was public and obvious. More subtle and covert was the role of AM broadcasting. On the AM wavebands the West's invective was more restrained, but even here there was a marked contrast and role reversal between the two sides. Baghdad Radio launched the 'Voice of Peace'—apparently modelled on Tokyo Radio's 'Zero Hour' of the Second World War. 'Voice of Peace' treated its listeners as intelligent people, offering good pop music interspersed with talk and comment. It reminded the GIs of US casualties in Vietnam, adding 'but this time it will be even bloodier—Iraq is a hell of a power'. Baghdad was substantiating what the troops had already been told by Western leaders. To any GI who had fought at Iwo Jima, the 'Voice of Peace' would have brought back chilling memories of the voice of Tokyo Rose: 'Be careful how you go to the latrines in the dark—there may be a Japanese soldier behind you.'

We may never know the full extent of Allied broadcasting using super powered transmitters in six locations: 4000 kW at Dubai, 2000 kW each at Qurayat, Jeddah, Damman and Batra in Egypt. These have been in place for over a decade. The USA has enough knowhow in all forms of covert broadcasting to 'cuddle up' to legitimate stations on the airwaves and then drown them out and assume their identity. This is not an easy task, involving changing frequency and modifying the mast radiator, but it is possible.

AM broadcasting comes into its own in a war zone, where television is of limited use. Whether in a bombed-out bazaar in Baghdad or a hastily improvised dugout in southern Iraq, the ubiquitous cheap battery-powered radio will be present. Through the eyes of CNN we saw peasants walking about with one ear cocked, quite possibly listening to the BBC or VOA.

AM broadcasting plays its most powerful role when it tries to bring about revolution in another country. The first time a radio transmitter was used for this aim was in the summer of 1918, by the USA, against Germany; the technique has also been used against Egypt (under Nasser), Persia and Iran, and Libya, to name only the local precedents, and we may be sure that similar efforts were made throughout the Gulf War.

Propaganda broadcasting has to be able to react swiftly to changing circumstances. As soon as the Gulf Crisis erupted, the BBC World Service deployed more of its units to the Middle East, using the flexibility of its computer-controlled transmitter stations. It also altered the style and content of its programmes to imitate the domestic BBC channel Radio 1, for the benefit of the UK soldiers deprived of their copies of The Sun.

Also visible in the Gulf War, as in most wars since the British fought Napoleon, was the propaganda leaflet. Evidently these archaic tools are still considered to have a role in destabilising the enemy forces—but as in the Second World War, the UK government ensured that none of the leaflets were reprinted by the media at home.

Leaping from 19th to 21st Century technology, the latest weapon for propaganda broadcasting is over-the-horizon (OTH) television. This is a combination of several disciplines in UHF transmission. As a propaganda weapon, it made its debut in 1990, when the USIA and VOA launched the controversial 'TV Marti', reaching the people of Cuba from a balloon 3000 m above the Florida Keys. Interestingly, the first country to invest in OTH technology on a large scale was Kuwait. Before the Iraqi invasion, a 30 000 kW UHF TV station was nearing completion on Failaka Island, Kuwait.

OTH TV is capable of reaching over 150 km, and often much further, with an acceptable signal. The technology is all-American. The basic element is the multistage depressed collector klystron and the klystrode, both products of Varian Associates. These tubes give high efficiencies, allowing them to be combined to give multiples of 70 kW up to 490 kW. The leading companies in UHF super power are Harris and Micro Communications Inc., another US company.

Merits of the different systems for television broadcasting

Broadcast band	Advantages	Disadvantages
Band III 174-230 MHz	Band III broadcasting is in use in most countries. TV set is cheap	Suitable for line-of-sight coverage with powers between 10-45 kW peak sync output
UHF bands IV and V 470-860 MHz	Band IV and V now the most popular television system. TV set is cheap	Suitable for line-of-sight coverage which can be extended higher power and tall masts
Satellite broadcasting using Ku-band (11 GHz)	Enables reception of TV signals over a large 'footprint' of a continent	Requires satellite antenna and converter as well as normal band IV-V set
As delivery systems for information/propaganda, all of these systems require the co-operation of the host country.		
Super-power broadcasting on bands IV and V	Requires only the use of a normal TV set	Super power broadcasting is the result of the latest technological developments. It uses ERP of the order of 20-30 MW. Developments still proceeding.
Super power television broadcasting cannot be intercepted or easily jammed. This will be one of the main delivery systems of the future for TV propaganda to neighbouring countries. First installation built on Failaka Island, Kuwait, in 1989 using 30 MW from very tall towers to reach adjoining territories.		

Merits of the different broadcast bands available for sound broadcasting, AM and FM

Broadcast band	Advantages	Disadvantages
AM broadcasting in the SW bands 3-30 MHz	Global coverage possible. Signals cannot easily be intercepted or jammed. The most popular choice for international broadcasters	Requires 300-500 kW transmitters, with sophisticated curtain array
AM broadcasting on the LW band 150-280 kHz	Does not require a sophisticated radio receiver, easy to tune into with very stable reception to 1500-3000 km	For long-distance broadcasting, requires transmitters with 1000-2000 kW carrier and tall, expensive antennas, 300 m high
AM broadcasting on the MW band 500-1605 kHz	Does not require a sophisticated radio receiver, easy to tune into. Night-time effect enables signals to travel much further, up to 4500 km with use of super power	For long-distance broadcasting, requires transmitters with 1000-2000 kW carrier and tall expensive antennas
FM broadcasting in the VHF band 88-106 MHz	Best suited for line-of-sight coverage, under 50 km. Good reception in primary area	Requires use of a 10 kW transmitter for best results. Of limited use only in information broadcasting. Ideal for short range, over-the-border application

Super-power broadcasting in the VHF-FM band is the latest technology. This uses effective radiated powers of 5-10 megawatts. First installations built on Failaka Island, Kuwait before Gulf war.

Appendix II

Sales of high-power transmitters since 1991

The effect of the political changes and upheavals in the world—the dismantling of the Berlin Wall, the collapse of the Soviet Union, the Gulf War and other events—has been to bring about a resurgence in propaganda broadcasting. The collapse of communism and the break-up of the Soviet Union was hailed as a victory by some Western nations. It was the culmination of 40 odd years of barrage broadcasting by BBC World Service, Voice of America and its more covert sister agencies Radio Free Europe and Radio Liberty. According to a report prepared by the US Presidential Task Force on US Government Information Broadcasting, 'the tax dollars spent were the most useful national security dollars spent in this century, they sent out words, ideas and they broke down a wall and broke down an empire'.

The Gulf War brought AM broadcasting on the short waves back into the headlines: SW broadcasts from London and Washington reached out to the citizens of war-torn Baghdad bringing allied news direct, uncensored by the Iraqi government. It required nothing more than the ubiquitous hand-held portable transistor radio. Moreover, unlike any other form of broadcasting, the SW receiver needs no high profile antenna to betray its presence.

Possibly inspired by the demonstrations of the effectiveness of SW broadcasting, a number of international broadcasters have announced further expansion plans for SW broadcasting in the high frequency bands. In December 1991 Radio France International embarked on a new project that will add another 17 SW transmitter systems to its transmission centres at Issoudun and Allouis, keeping to the French policy of having its powerful transmitter complexes on French soil. Voice of America, on the other hand, is still in possession of transmitting stations around the world, many in countries with repressive right-wing regimes and the VOA expansion goes ahead as if the Cold War had never ended.

This resurgence in information broadcasting in the HF spectrum is clearly shown in my statistical analysis of transmitter sales since January 1991, and comparing same with the similar statistical analysis (Table 32.5) for the five-year period up to December 1990. Between 1985 and 1990 a total of 200 transmitters were sold by the big five manufacturers. This figure excluded sales for transmitters with less than 100 kW carrier power.

From January 1991 to June 1992 a total of 134 transmitters were sold or on order. All figures are from manufacturers' lists or from personal sources and are subject to minor discrepancies.

Company	Units sold	Breakdown
ABB	21	Three 500 kW, 15 250 kW, one 600 kW, two 100 kW transmitters
AEG †	6	Probably 500 kW transmitters
Continental	19	Nine 500 kW, ten less than 500 kW transmitters
Harris	11	Ten 100 kW, one 300 kW, all new DX series transmitters
RIZ	20	All 100 kW, mostly transportables sold to Middle East
Thomson CSF	43	Two 1000 kW, 23 500 kW, 18 100-300 kW transmitters
Marconi	15	Various powers up to 500 kW

Sales figures are divided between Western broadcasting agencies and the Arab countries in the Middle East, Gulf States and North Africa. It is impossible to estimate growth since 1991 accurately because the first sales analysis referred to a five-year period up to December 1990, whereas the second is for an 18-month period, and this rate may not continue for another five years. Even so, the trend points to a quite startling growth rate in information broadcasting.

Two other facts that emerge from the latest statistical analysis are the numbers of transmitter manufacturers catering for the high power market—now increased to seven—and the league table measured by sales. Up to 1990, it was the Swiss company ABB* that held the number one position for five years. Now the French company Thomson-CSF is dominating the market with 43 high-power transmitters sold in the past 18 months. In the late 1970s to the early 1980s, it was Thomson that dominated the global market, supplying three out of every five transmitters in the world. It looks as if this achievement could be repeated in the late 1990s.

However, ABB is still considered by many to be a world leader in high-power transmission sciences, and if a consensus were taken the two companies would be finely balanced. Both are world leaders and have a high capability in executing large installations, on a turn-key basis, in environmentally hostile parts of the world.

†Now Telefunken Sendertechnik.

*In 1993 the Swiss/Swedish giant ABB merged its transmitter manufacturing business with that of Thomson-CSF of France, creating the world's largest broadcasting transmission manufacturer Thomcast.

Appendix III

Low-profile transmitters

Following intense international competition, Thomson-CSF has won the contract to renew Telediffusion de France's (TDF) high power shortwave (SW) transmission system. TDF's system, which transmits Radio France International (RFI), will be expanded and partly replaced by Thomson's new SW transmitting modules called ALLISS.

Continued developments in massive high gain HF curtain arrays have seen the construction of huge systems that cover the broadcast band from 4–26 MHz, with a total length of 300–350 m required for each transmitter. HF transmitting stations like those at Rampisham (BBC World Service), Greenville USA (Voice of America), Wertachel (Deutsche Welle) and Allouis/Issoudun (RFI) have become more complex in construction. These stations also demand more real estate and in other aspects approach the limits of physics in the switching systems that are needed to connect transmitters to antennas.

The concept of Thomson's TDF project is revolutionary in the levels of RF power being handled. It has reversed all existing notions about transmitting station design. The transmitting hall has gone along with the even larger structure needed to house the giant switching matrix which is often the major cause of unreliability. Gone also is the 100 km (plus) of transmission lines (with attendant problems of cross modulation) and the antenna farm, which often exceeded 300 acres of land.

In their place, Thomson has introduced a new concept. This is a single 500 kW SW transmitter in a silo, surmounted by a rotatable support structure which carries a dynamic balance of back-to-back curtains. A central screen of wires prevents back radiation from the array in use. This concept, claimed by Thomson to reduce transmitting station costs by as much as 30 percent, has other important advantages. It does away with the need for standby transmitters and standby antennas—for the simple reason that one transmitter can focus its fire power onto any azimuth bearing and so broadcast to any part of the world.

ALLISS derives its name from Allouis and Issoudun, RFI's transmitter sites where 12 × 100 kW and 8 × 500 kW transmitters are installed along with 88 curtain arrays and a massive switching matrix. As part of a re-investment programme the 8 × 100 kW transmitter will be taken out of service and replaced

This appendix is from: WOOD, J.: 'New wave at RFI', *International Broadcasting*, Jan/Feb 1992, p. 36. Reproduced by permission of International Thomson Business Publishing.

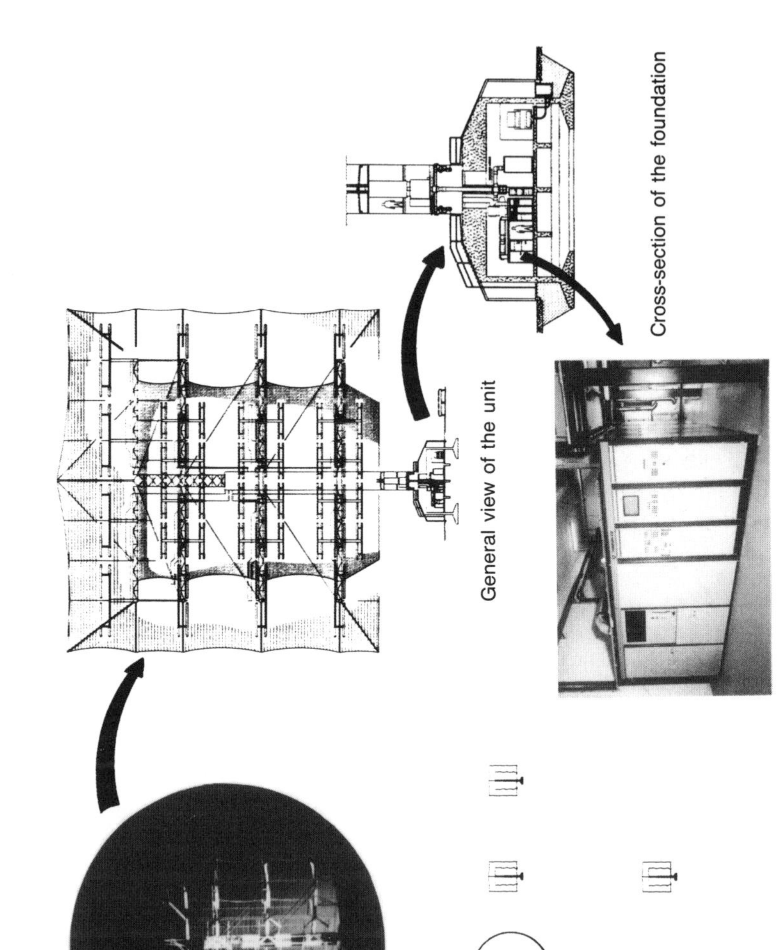

Figure A.1 *The Thomson ALLISS SW centre*

with 15 ALLISS installations. These will be dispersed in order to eliminate problems with feeder interaction and unwanted mutual coupling between antennas. The advantages of this configuration are considerable. Apart from the obvious technical ones, the concept offers greater immunity to system failure and from sabotage or direct attack from the air. It also offers a solution to the growing ecological problems of massive amounts of electromagnetic radiation in a small area.

This new approach to station design has been a well-kept secret. The company's Guy Noel le Carvennec merely hinted to me in 1989 that the company was working on something new: 'The modern SW transmitter has evolved to a point where it is other components in the transmission system that are the limiting factor... We are now turning our attention to these things'.

The ALLISS concept is the outcome. If Thomson ever had any doubts about the direction of its development programme then these would surely have been removed in February 1991 during the Gulf War, when allied aircraft disabled and destroyed Iraq's SW complex at Balad. This was the largest SW station in the world, yet all it needed to put it out of action was a direct hit on the antenna switching matrix. If this station had been constructed on the ALLISS principle, it would have been difficult to disable.

The ALLISS concept seems to be the ideal solution for less stable parts of the world such as the Middle East. However, it also offers powerful advantages of a geopolitical nature to broadcasters who need to have a rapid reaction to fast moving and unpredictable political events. In such circumstances, governments sometimes react with barrage broadcasting, which is the technique of switching as many transmitters as possible to a particular target area.

Barrage broadcasting played a powerful role during the Cold War but it was never as successful as it could have been, because the fixed curtain arrays were not always on the desired azimuth bearing. Even during the worst phases of the Cold War, the BBC never managed to get more than about one third of its total transmission power to the target zone.

With the ALLISS concept, every one of the 500 kW SW transmitters can direct its fire power onto the target zone in less than three minutes, sometimes less. This is all the time needed to rotate the high gain curtain array on its central column to any desired angle of bearing. With the Cold War now a thing of the past, who is to say that there will never be another international crisis—perhaps within the USSR itself, or in the Middle East. This is why international broadcasters are at this very time planning even greater expansion programmes in their broadcasting power.

Index